Applied Colloid
and
Surface Chemistry

Applied Colloid
and
Surface Chemistry

Richard M. Pashley and **Marilyn E. Karaman**

Department of Chemistry, The National University of Australia, Canberra, Australia

John Wiley & Sons, Ltd

OCM 56420248

Other Wiley Editorial Offices

John Wiley & Sons Inc., 111 River Street, Hoboken, NJ 07030, USA

Jossey-Bass, 989 Market Street, San Francisco, CA 94103-1741, USA

Wiley-VCH Verlag GmbH, Boschstr. 12, D-69469 Weinheim, Germany

John Wiley & Sons Australia Ltd, 33 Park Road, Milton, Queensland 4064, Australia

John Wiley & Sons (Asia) Pte Ltd, 2 Clementi Loop # 02-01, Jin Xing Distripark, Singapore
129809

John Wiley & Sons Canada Ltd, 22 Worcester Road, Etobicoke, Ontario, Canada M9W 1L1

Wiley also publishes its books in a variety of electronic formats. Some content that appears in
print may not be available in electronic books.

Library of Congress Cataloging-in-Publication Data

Pashley, Richard M.
    Applied colloid and surface chemistry / Richard M. Pashley and Marilyn E. Karaman.
      p.   cm.
    Includes bibliographical references and index.
    ISBN 0 470 86882 1 (cloth : alk. paper) — ISBN 0 470 86883 X (pbk. : alk. paper)
  1. Colloids. 2. Surface chemistry. I. Karaman, Marilyn E.   II. Title.
    QD549.P275 2004
    541'.345 — dc22

                                          2004020586

British Library Cataloguing in Publication Data
A catalogue record for this book is available from the British Library

ISBN 0 470 86882 1 Hardback
    0 470 86883 X Paperback

Typeset in 11/13½pt Sabon by SNP Best-set Typesetter Ltd., Hong Kong
Printed and bound in Great Britain by TJ International Ltd, Padstow, Cornwall
This book is printed on acid-free paper responsibly manufactured from sustainable forestry in
which at least two trees are planted for each one used for paper production.

Those that can, teach

Sit down before fact as a little child, be prepared
to give up every preconceived notion, follow
humbly wherever and to whatever abysses nature
leads, or you shall learn nothing.

Thomas Henry Huxley (1860)

Contents

Preface

This book was written following several years of teaching this material to third-year undergraduate and honours students in the Department of Chemistry at the Australian National University in Canberra, Australia. Science students are increasingly interested in the application of their studies to the real world and colloid and surface chemistry is an area that offers many opportunities to apply learned understanding to everyday and industrial examples. There is a lack of resource materials with this focus and so we have produced the first edition of this book. The book is intended to take chemistry or physics students with no background in the area, to the level where they are able to understand many natural phenomena and industrial processes, and are able to consider potential areas of new research. Colloid and surface chemistry spans the very practical to the very theoretical, and less mathematical students may wish to skip some of the more involved derivations. However, they should be able to do this and still maintain a good basic understanding of the fundamental principles involved. It should be remembered that a thorough knowledge of theory can act as a barrier to progress, through the inhibition of further investigation. Students asking ignorant but intelligent questions can often stimulate valuable new research areas.

The book contains some recommended experiments which we have found work well and stimulate students to consider both the fundamental theory and industrial applications. Sample questions have also been included in some sections, with detailed answers available on our web site.

Although the text has been primarily aimed at students, researchers in cognate areas may also find some of the topics stimulating. A reasonable background in chemistry or physics is all that is required.

1

Introduction

Introduction to the nature of colloids and the linkage between colloids and surface properties. The importance of size and surface area. Introduction to wetting and the industrial importance of surface modifications.

Introduction to the nature of colloidal solutions

The difference between macroscopic and microscopic objects is clear from everyday experience. For example, a glass marble will sink rapidly in water; however, if we grind it into sub-micron-sized particles, these will float or disperse freely in water, producing a visibly cloudy 'solution', which can remain stable for hours or days. In this process we have, in fact, produced a 'colloidal' dispersion or solution. This dispersion of one (finely divided or microscopic) phase in another is quite different from the molecular mixtures or 'true' solutions formed when we dissolve ethanol or common salt in water. Microscopic particles of one phase dispersed in another are generally called *colloidal solutions* or *dispersions*. Both nature and industry have found many uses for this type of solution. We will see later that the properties of colloidal solu-

Applied Colloid and Surface Chemistry Richard M. Pashley and Marilyn E. Karaman
© 2004 John Wiley & Sons, Ltd.   ISBN 0 470 86882 1 (HB)   0 470 86883 X (PB)

tions are intimately linked to the high surface area of the dispersed phase, as well as to the chemical nature of the particle's surface.

> Historical note: The term 'colloid' is derived from the Greek word 'kolla' for glue. It was originally used for gelatinous polymer colloids, which were identified by Thomas Graham in 1860 in experiments on osmosis and diffusion.

It turns out to be very useful to dissolve (or more strictly disperse) solids, such as minerals and metals, in water. But how does it happen? We can see why from simple physics. Three fundamental forces operate on fine particles in solution:

(1) a gravitational force, tending to settle or raise particles depending on their density relative to the solvent;

(2) a viscous drag force, which arises as a resistance to motion, since the fluid has to be forced apart as the particle moves through it;

(3) the 'natural' kinetic energy of particles and molecules, which causes Brownian motion.

If we consider the first two forces, we can easily calculate the terminal or limiting velocity, V, (for settling or rising, depending on the particle's density relative to water) of a spherical particle of radius r. Under these conditions, the viscous drag force must equal the gravitational force. Thus, at a settling velocity, V, the viscous drag force is given by: $F_{drag} = 6\pi r V\eta = 4\pi r^3 g(\rho_p - \rho_w)/3 = F_{gravity}$, the gravitational force, where η is the viscosity of water and the density difference between particle and water is $(\rho_p - \rho_w)$. Hence, if we assume a particle–water density difference of $+1\,g\,cm^{-3}$, we obtain the results:

r (Å)	100	1000	10 000	10^5	10^6
r (µm)	0.01	0.1	1	10	100
V (cm s^{-1})	2×10^{-8}	2×10^{-6}	2×10^{-4}	2×10^{-2}	2

Clearly, from factors (1) and (2), small particles will take a very long time to settle and so a fine dispersion will be stable almost indefinitely, even for materials denser than water. But what of factor (3)? Each particle, independent of size, will have a kinetic energy, on average, of around $1kT$. So the typical, random speed (v) of a particle (in any direction) will be roughly given by:

$$mv^2/2 \cong 1kT \cong 4 \times 10^{-21}\,\text{J} \quad \text{(at room temperature)}$$

Again, if we assume that $\rho_p = 2\,\text{g cm}^{-3}$, then we obtain the results:

r (Å)	100	1000	10 000	10^5	10^6
r (μm)	0.01	0.1	1	10	100
v (cm s^{-1})	10^2	3	0.1	3×10^{-3}	1×10^{-4}

These values suggest that kinetic random motion will dominate the behaviour of small particles, which will not settle and the dispersion will be completely stable. However, this point is really the beginning of 'colloid science'. Since these small particles have this kinetic energy they will, of course, collide with other particles in the dispersion, with collision energies ranging up to at least $10kT$ (since there will actually be a distribution of kinetic energies). If there are attractive forces between the particles – as is reasonable since most colloids were initially formed via a vigorous mechanical process of disruption of a macroscopic or large body – each collision might cause the growth of large aggregates, which will then, for the reasons already given, settle out, and we will no longer have a stable dispersion! The colloidal solution will coagulate and produce a solid precipitate at the bottom of a clear solution.

There is, in fact, a ubiquitous force in nature, called the *van der Waals force* (vdW), which is one of the main forces acting between molecules and is responsible for holding together many condensed phases, such as solid and liquid hydrocarbons and polymers. It is responsible for about one third of the attractive force holding liquid water molecules together. This force was actually first observed as a correction to the ideal gas equation and is attractive even between neutral gas molecules, such as oxygen and nitrogen, in a vacuum. Although electromagnetic in origin (as we will see later), it is much weaker than the Coulombic force acting between ions.

The forces involved in colloidal stability

Although van der Waals forces will always act to coagulate dispersed colloids, it is possible to generate an opposing repulsive force of comparable strength. This force arises because most materials, when dispersed in water, ionize to some degree or selectively adsorb ions from solution and hence become charged. Two similarly charged colloids will repel each other via an electrostatic repulsion, which will oppose coagulation. The stability of a colloidal solution is therefore critically dependent on the charge generated at the surface of the particles. The combination of these two forces, attractive van der Waals and repulsive electrostatic forces, forms the fundamental basis for our understanding of the behaviour and stability of colloidal solutions. The corresponding theory is referred to as the DLVO (after Derjaguin, Landau, Verwey and Overbeek) theory of colloid stability, which we will consider in greater detail later. The stability of any colloidal dispersion is thus determined by the behaviour of the surface of the particle via its surface charge and its short-range attractive van der Waals force.

Our understanding of these forces has led to our ability to selectively control the electrostatic repulsion, and so create a powerful mechanism for controlling the properties of colloidal solutions. As an example, if we have a valuable mineral embedded in a quartz rock, grinding the rock will both separate out pure, individual quartz and the mineral particles, which can both be dispersed in water. The valuable mineral can then be selectively coagulated, whilst leaving the unwanted quartz in solution. This process is used widely in the mining industry as the first stage of mineral separation. The alternative of chemical processing, for example, by dissolving the quartz in hydrofluoric acid, would be both expensive and environmentally unfriendly.

It should be realized, at the outset, that colloidal solutions (unlike true solutions) will almost always be in a metastable state. That is, an electrostatic repulsion prevents the particles from combining into their most thermodynamically stable state, of aggregation into the macroscopic form, from which the colloidal dispersion was (artificially) created in the first place. On drying, colloidal particles will often remain separated by these repulsive forces, as illustrated by Figure 1.1, which shows a scanning electron microscope picture of mono-disperse silica colloids.

Figure 1.1 Scanning electron microscope image of dried, mono-disperse silica colloids.

Types of colloidal systems

The term 'colloid' usually refers to particles in the size range 50 Å to 50 μm but this, of course, is somewhat arbitrary. For example, blood could be considered as a colloidal solution in which large blood cells are dispersed in water. Often we are interested in solid dispersions in aqueous solution but many other situations are also of interest and industrial importance. Some examples are given in Table 1.1.

Table 1.1

Dispersed phase	Dispersion medium	Name	Examples
Liquid	Gas	Liquid aerosol	Fogs, sprays
Solid	Gas	Solid aerosol	Smoke, dust
Gas	Liquid	Foam	Foams
Liquid	Liquid	Emulsion	Milk, Mayonnaise
Solid	Liquid	'Sol' or colloidal solution	Au sol, AgI sol
		Paste at high concentration	Toothpaste
Gas	Solid	Solid foam	Expanded polystyrene
Liquid	Solid	Solid emulsion	Opal, pearl
Solid	Solid	Solid suspension	Pigmented plastics

The properties of colloidal dispersions are intimately linked to the high surface area of the dispersed phase and the chemistry of these interfaces. This linkage is well illustrated by the titles of two of the main journals in this area: the *Journal of Colloid and Interface Science* and *Colloids and Surfaces*. The natural combination of colloid and surface chemistry represents a major area of both research activity and industrial development. It has been estimated that something like 20 per cent of all chemists in industry work in this area.

The link between colloids and surfaces

The link between colloids and surfaces follows naturally from the fact that particulate matter has a high surface area to mass ratio. The surface area of a 1 cm diameter sphere ($4\pi r^2$) is 3.14 cm^2, whereas the surface area of the same amount of material but in the form of 0.1 µm diameter spheres (i.e. the size of the particles in latex paint) is 314 000 cm^2. The enormous difference in surface area is one of the reasons why the properties of the surface become very important for colloidal solutions. One everyday example is that organic dye molecules or pollutants can be effectively removed from water by adsorption onto particulate activated charcoal because of its high surface area. This process is widely used for water purification and in the oral treatment of poison victims.

Although it is easy to see that surface properties will determine the stability of colloidal dispersions, it is not so obvious why this can also

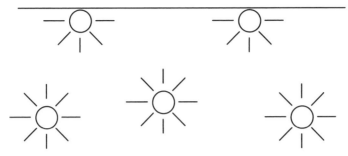

Figure 1.2 Schematic diagram to illustrate the complete bonding of liquid molecules in the bulk phase but not at the surface.

Table 1.2

Liquid	Surface energy in mJ m^{-2} (at 20 °C)	Type of intermolecular bonding
Mercury	485	metallic
Water	72.8	hydrogen bonding + vdW
n-Octanol	27.5	hydrogen bonding + vdW
n-Hexane	18.4	vdW
Perfluoro-octane	12	weak vdW

be the case for some properties of macroscopic objects. As one important illustration, consider Figure 1.2, which illustrates the interface between a liquid and its vapour. Molecules in the bulk of the liquid can interact via attractive forces (e.g. van der Waals) with a larger number of nearest neighbours than those at the surface. The molecules at the surface must therefore have a higher energy than those in bulk, since they are partially freed from bonding with neighbouring molecules. Thus, work must be done to take fully interacting molecules from the bulk of the liquid to create any new surface. This work gives rise to the surface energy or tension of a liquid. Hence, the stronger the intermolecular forces between the liquid molecules, the greater will this work be, as is illustrated in Table 1.2.

The influence of this surface energy can also be clearly seen on the macroscopic shape of liquid droplets, which in the absence of all other forces will always form a shape of minimum surface area – that is, a sphere in a gravity-free system. This is the reason why small mercury droplets are always spherical.

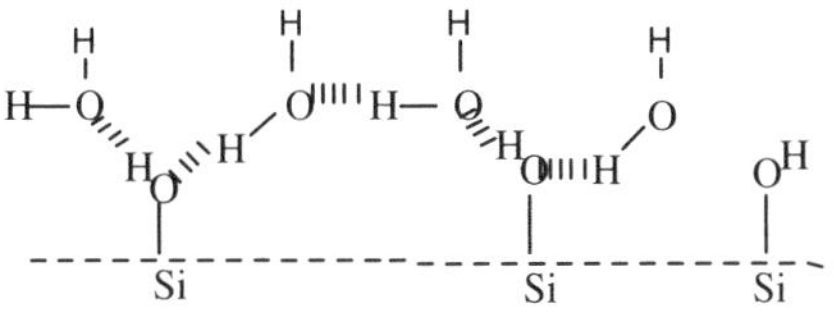

Figure 1.3 Water molecules form hydrogen bonds with the silanol groups at the surface of clean glass.

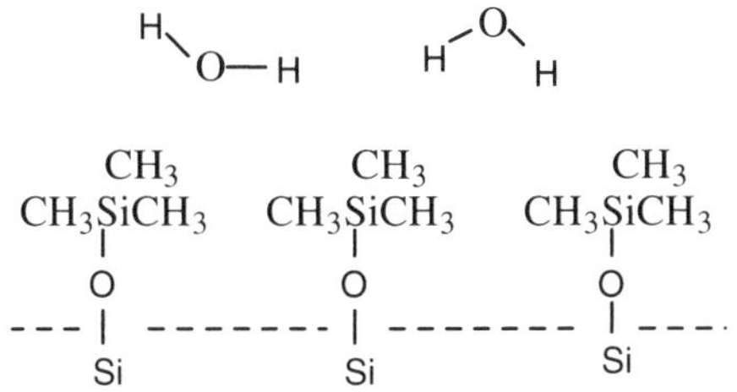

Figure 1.4 Water molecules can only weakly interact (by vdw forces) with a methylated glass surface.

Wetting properties and their industrial importance

Although a liquid will always try to form a minimum-surface-area shape, if no other forces are involved, it can also interact with other macroscopic objects, to reduce its surface tension via molecular bonding to another material, such as a suitable solid. Indeed, it may be energetically favourable for the liquid to interact and 'wet' another material. The wetting properties of a liquid on a particular solid are very important in many everyday activities and are determined solely by surface properties. One important and common example is that of water on clean glass. Water wets clean glass (Figure 1.3) because of the favourable hydrogen bond interaction between the surface silanol groups on glass and adjacent water molecules.

However, exposure of glass to Me_3SiCl vapour rapidly produces a 0.5 nm layer of methyl groups on the surface. These groups cannot hydrogen-bond and hence water now does not wet and instead forms high 'contact angle' (θ) droplets and the glass now appears to be hydrophobic, with water droplet beads similar to those observed on paraffin wax (Figure 1.5).

This dramatic macroscopic difference in wetting behaviour is caused by only a thin molecular layer on the surface of glass and clearly demonstrates the importance of surface properties. The same type of

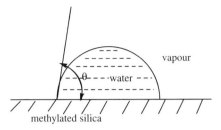

Figure 1.5 A non-wetting water droplet on the surface of methylated, hydrophobic silica.

effect occurs every day, when dirty fingers coat grease onto a drinking glass! Surface treatments offer a remarkably efficient method for the control of macroscopic properties of materials. When insecticides are sprayed onto plant leaves, it is vital that the liquid wet and spread over the surface. Another important example is the froth flotation technique, used by industry to separate about a billion tons of ore each year. Whether valuable mineral particles will attach to rising bubbles and be 'collected' in the flotation process, is determined entirely by the surface properties or surface chemistry of the mineral particle, and this can be controlled by the use of low levels of 'surface-active' materials, which will selectively adsorb and change the surface properties of the mineral particles. Very large quantities of minerals are separated simply by the adjustment of their surface properties.

Although it is relatively easy to understand why some of the macroscopic properties of liquids, especially their shape, can depend on surface properties, it is not so obvious for solids. However, the strength of a solid is determined by the ease with which micro-cracks propagate, when placed under stress, and this depends on its surface energy, that is the amount of (surface) work required to continue the crack and hence expose new surface. This has the direct effect that materials are stronger in a vacuum, where their surface energy is not reduced by the adsorption of either gases or liquids, typically available under atmospheric conditions.

Many other industrial examples where colloid and surface chemistry plays a significant role will be discussed later, these include:

- latex paint technology
- photographic emulsions
- soil science
- soaps and detergents

- food science
- mineral processing.

Recommended resource books

Adamson, A.W. (1990) *Physical Chemistry of Surfaces*, 5th edn, Wiley, New York

Birdi, K.S. (ed.) (1997) *CRC Handbook of Surface and Colloid Chemistry*, CRC Press, Boca Raton, FL

Evans, D.F. and Wennerstrom, H. (1999) *The Colloidal Domain*, 2nd edn, Wiley, New York

Hiemenz, P.C. (1997) *Principles of Colloid and Surface Chemistry*, 3rd edn, Marcel Dekker, New York

Hunter, R.J. (1987) *Foundations of Colloid Science, Vol. 1*, Clarendon Press, Oxford

Hunter, R.J. (1993) *Introduction to Modern Colloid Science*, Oxford Sci. Publ., Oxford

Israelachvili, J.N. (1985) *Intermolecular and Surface Forces*, Academic Press, London

Shaw, D.J. (1992) *Introduction to Colloid and Surface Chemistry*, 4th edn, Butterworth-Heinemann, Oxford, Boston

Appendices

A Some historical notes on colloid and surface chemistry

Robert Hooke (1661) investigates capillary rise.

John Freind at Oxford (1675–1728) was the first person to realize that inter-molecular forces are of shorter range than gravity.

Young (1805) estimated range of intermolecular forces at about 0.2 nm. Turns out to be something of an underestimate.

Young and Laplace (1805) derived meniscus curvature equation.

Brown (1827) observed the motion of fine particles in water.

Van der Waals (1837–1923) was a schoolmaster who produced a doctoral thesis on the effects of intermolecular forces on the properties of gases (1873).

Graham (1860) had recognized the existence of colloids in the mid 19th century.

Faraday (1857) made colloidal solutions of gold.

Schulze and Hardy (1882–1900) studied the effects of electrolytes on colloid stability.

Perrin (1903) used terms 'lyophobic' and 'lyophilic' to denote irreversible and reversible coagulation.

Ostwald (1907) developed the concepts of 'disperse phase' and 'dispersion medium'.

Gouy and Chapman (1910–13) independently used the Poisson–Boltzmann equations to describe the diffuse electrical double-layer formed at the interface between a charged surface and an aqueous solution.

Ellis and Powis (1912–15) introduced the concept of the critical zeta potential for the coagulation of colloidal solutions.

Fritz London (1920) first developed a theoretical basis for the origin of inter-molecular forces.

Debye (1920) used polarizability of molecules to estimate attractive forces.

Debye and Hückel (1923) used a similar approach to Gouy and Chapman to calculate the activity coefficients of electrolytes.

Stern (1924) introduced the concept of specific ion adsorption at surfaces.

Kallmann and Willstätter (1932) calculated van der Waals force between colloidal particles using the summation procedure and suggested that a complete picture of colloid stability could be obtained on the basis of electrostatic double-layer and van der Waals forces.

Bradley (1932) independently calculated van der Waals forces between colloidal particles.

Hamaker (1932) and **de Boer** (1936) calculated van der Waals forces between macroscopic bodies using the summation method.

Derjaguin and Landau, and Verwey and Overbeek (1941–8) developed the DLVO theory of colloid stability.

Lifshitz (1955–60) developed a complete quantum electrodynamic (continuum) theory for the van der Waals interaction between macroscopic bodies.

B Dispersed particle sizes

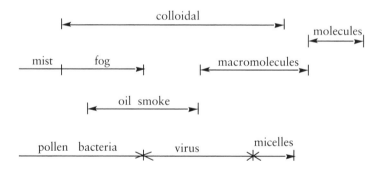

2
Surface Tension and Wetting

The equivalence of the force and energy description of surface tension and surface energy. Derivation of the Laplace pressure and a description of common methods for determining the surface tension of liquids. The surface energy and cohesion of solids, liquid wetting and the liquid contact angle. Laboratory projects for measuring the surface tension of liquids and liquid contact angles.

The equivalence of the force and energy description of surface tension and surface energy

It is easy to demonstrate that the surface energy of a liquid actually gives rise to a 'surface tension' or force acting to oppose any increase in surface area. Thus, we have to 'blow' to create a soap bubble by stretching a soap film. A spherical soap bubble is formed in response to the tension in the bubble surface (Figure 2.1). The soap film shows interference colours at the upper surface, where the film is starting to thin, under the action of gravity, to thicknesses of the order of the wavelength of light. Some beautiful photographs of various types of soap films are given in *The Science of Soap Films and Soap Bubbles* by C. Isenberg (1992).

Applied Colloid and Surface Chemistry Richard M. Pashley and Marilyn E. Karaman
© 2004 John Wiley & Sons, Ltd.   ISBN 0 470 86882 1 (HB)   0 470 86883 X (PB)

Figure 2.1 Photograph of a soap bubble.

If we stretch a soap film on a wire frame, we find that we need to apply a significant, measurable force, F, to prevent collapse of the film (Figure 2.2). The magnitude of this force can be obtained by consideration of the energy change involved in an infinitesimal movement of the cross-bar by a distance dx, which can be achieved by doing reversible work on the system, thus raising its free energy by a small amount Fdx. If the system is at equilibrium, this change in (free) energy must be exactly equal to the increase in surface (free) energy $(2dxl\gamma)$ associated with increasing the area of both surfaces of the soap film. Hence, at equilibrium:

$$Fdx = 2dxl\gamma \tag{2.1}$$

or

$$\gamma = F/2l \tag{2.2}$$

It is precisely this, that work has to be done to increase a liquid's surface area, that makes the surface of a liquid behave like a stretched skin,

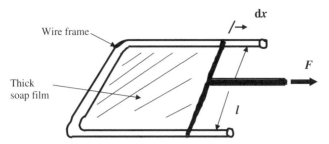

Figure 2.2 Diagram of a soap film stretched on a wire frame.

hence the term 'surface tension'. It is this tension that allows a water boatman insect to travel freely on the surface of a pond, locally deforming the skin-like surface of the water.

This simple experimental system clearly demonstrates the equivalence of surface energy and tension. The dimensions of surface energy, $mJ\,m^{-2}$, are equivalent to those of surface tension, $mN\,m^{-1}$. For pure water, an energy of about $73\,mJ$ is required to create a $1\,m^2$ area of new surface. Assuming that one water molecule occupies an area of roughly $12\,Å^2$, the free energy of transfer of one molecule of water from bulk to the surface is about $3\,kT$ (i.e. $1.2 \times 10^{-20}\,J$), which compares with roughly $8\,kT$ per hydrogen bond. The energy or work required to create new water–air surface is so crucial to a newborn baby that nature has developed lung surfactants specially to reduce this work by about a factor of three. Premature babies often lack this surfactant and it has to be sprayed into their lungs to help them breathe.

Derivation of the Laplace pressure equation

Since it is relatively easy to transfer molecules from bulk liquid to the surface (e.g. shake or break up a droplet of water), the work done in this process can be measured and hence we can obtain the value of the surface energy of the liquid. This is, however, obviously not the case for solids (see later section). The diverse methods for measuring surface and interfacial energies of liquids generally depend on measuring either the pressure difference across a curved interface or the equilibrium (reversible) force required to extend the area of a surface, as above. The former method uses a fundamental equation for the pressure generated across any curved interface, namely the Laplace equation, which is derived in the following section.

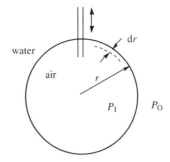

Figure 2.3   Diagram of a spherical air bubble in water.

Let us consider the conditions under which an air bubble (i.e. a curved surface) is stable. Consider the case of an air bubble produced in water by blowing through a tube (Figure 2.3). Obviously, to blow the air bubble we must have applied a higher pressure, P_I, inside the bubble, compared with the external pressure in the surrounding water (P_O). The bubble will be stable when there is no net air flow, in or out, and the bubble radius stays constant. Under these, equilibrium, conditions there will be no free energy change in the system for any infinitesimal change in the bubble radius, that is, $dG/dr = 0$, where dr is an infinitesimal decrease in bubble radius. If the bubble were to collapse by a small amount dr, the surface area of the bubble will be reduced, giving a decrease in the surface free energy of the system. The only mechanism by which this change can be prevented is to raise the pressure inside the bubble so that $P_I > P_O$ and work has to be done to reduce the bubble size. The bubble will be precisely at equilibrium when the change in free energy due to a reduced surface area is balanced by this work. For an infinitesimal change, dr, the corresponding free energy change of this system is given by the sum of the decrease in surface free energy and the mechanical work done against the pressure difference across the bubble surface, thus:

$$dG = -\gamma\left\{4\pi r^2 - 4\pi(r - dr)^2\right\} + (P_I - P_O)4\pi r^2 dr \qquad (2.3)$$

$$= -8\pi r dr\gamma + \Delta P 4\pi r^2 dr \qquad (2.4)$$

(ignoring higher-order terms). At equilibrium $dG/dr = 0$ and hence:

$$-8\pi r\gamma + \Delta P 4\pi r^2 = 0 \qquad (2.5)$$

$$\therefore \quad \Delta P = \frac{2\gamma}{r} \qquad (2.6)$$

This result is the Laplace equation for a single, spherical interface. In general, that is for any curved interface, this relationship expands to include the two principal radii of curvature, R_1 and R_2:

$$\Delta P = \gamma \left(\frac{1}{R_1} + \frac{1}{R_2} \right) \qquad (2.7)$$

Note that for a spherical surface $R_1 = R_2 = r$ and we again obtain (2.6). This equation is sometimes referred to as the 'Young–Laplace equation'. The work required to stretch the rubber of a balloon is directly analogous to the interfacial tension of the liquid surface. That the pressure inside a curved meniscus must be greater than that outside is most easily understood for gas bubbles (and balloons) but is equally valid for liquid droplets. The Laplace equation is also useful in calculating the initial pressure required to nucleate very small bubbles in liquids. Very high internal pressures are required to nucleate small bubbles and this remains an issue for de-gassing, boiling and decompression sickness. Some typical values for bubbles in water are:

Bubble radius in nm	1	2	10	1000
Laplace pressure in bar	1440	720	144	1.44

The high pressures associated with high-curvature interfaces leads directly to the use of boiling chips to help nucleate bubbles with lower curvature using the porous, angular nature of the chips (Figure 2.4).

Methods for determining the surface tension of liquids

The equilibrium curvature of a liquid surface or meniscus depends not just on its surface tension but also on its density and the effect of gravity. The variation in curvature of a meniscus surface must be due to hydrostatic pressure differences at different vertical points on the meniscus. If the curvature at a given starting point on a surface is known, the adjacent curvature can be obtained from the Laplace equation and its change in hydrostatic pressure $\Delta h \rho g$. In practice the liquid

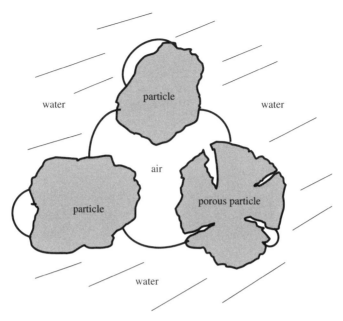

Figure 2.4 Schematic illustration of particles (e.g. boiling chips) used to reduce air bubble curvature.

droplet, say in air, has a constant volume and is physically constrained at some point, for example when a pendant drop is constrained by the edge of a capillary tube (Figure 2.5). For given values of the total volume, the radius of the tube R, the density ρ and the surface energy γ, the shape of the droplet is completely defined and can be calculated using numerical methods (e.g. the Runge–Kutta method) to solve the Laplace equation. Beautiful shapes can be generated using this numerical procedure. Although a wide variety of shapes can be generated using the Laplace equation in a gravitation field, only those shapes which give a minimum in the total energy (that is, surface and potential) will be physically possible. In practice, a continuous series of numerically generated profiles are calculated until the minimum energy shape is obtained.

It is interesting to consider the size of droplets for which surface (tension) forces, compared with gravity, dominate liquid shapes. A simple balance of these forces is given by the relation:

$$\text{length} = \sqrt{\frac{\gamma}{\rho g}} \approx 4\,\text{mm}, \quad \text{for water}$$

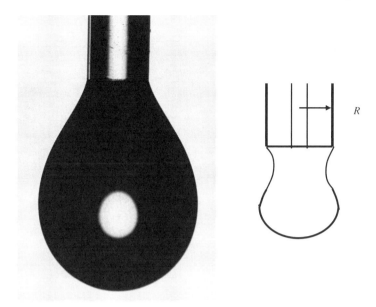

Figure 2.5 Photograph and diagram of a pendant liquid drop at the end of a glass capillary tube.

Thus we would expect water to 'climb' up the walls of a clean (i.e. water-wetting) glass vessel for a few millimetres but not more, and we would expect a sessile water droplet to reach a height of several mm on a hydrophobic surface, before the droplet surface is flattened by gravitational forces. The curved liquid border at the perimeter of a liquid surface or film is called the 'Plateau border' after the French scientist who studied liquid shapes after the onset of blindness, following his personal experiments on the effects of sunlight on the human eye.

The observation of a pendant drop is one of the best methods of measuring surface and interfacial energies of liquids. Either the drop can be photographed and the profile digitized or published tables can be used to obtain γ from only the drop volume and the minimum and maximum widths of the drop. Another simple method of measuring the surface energy of liquids is using a capillary tube. In this method the height to which the liquid rises, in the capillary, above the free liquid surface is measured. This situation is illustrated in Figure 2.6. Using the Laplace equation the pressure difference between points A and B is given simply by $\Delta P = 2\gamma/r$, if we assume that the meniscus is hemispherical and of radius r. However, this will be accurate only if the liquid wets the walls of the glass tube. If the liquid has a finite contact

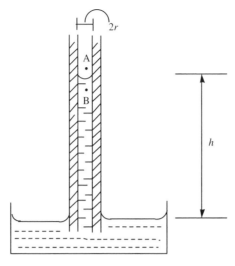

Figure 2.6 Schematic diagram of the rise of a liquid that wets the inside walls of a capillary tube.

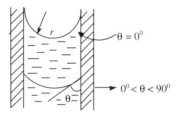

Figure 2.7 Schematic diagram of the shape of a meniscus for wetting and non-wetting liquids.

angle θ with the glass as in Figure 2.7, then from simple geometry (again assuming the meniscus is spherical)

$$\Delta P = \frac{2\gamma \cos\theta}{r} \tag{2.8}$$

Note that if $\theta > 90°$ (e.g. mercury on glass), the liquid will actually fall below the reservoir level and the meniscus will be curved in the opposite direction.

The pressure difference between points A and B must be equal to the hydrostatic pressure difference $h\rho g$ (where ρ is the density of the liquid and the density of air is ignored). Thus, we obtain the result that

$$hpg = \frac{2\gamma \cos\theta}{r} \qquad (2.9)$$

and hence

$$\gamma = \frac{rhpg}{2\cos\theta} \qquad (2.10)$$

from which measurement of the capillary rise and the contact angle gives the surface tension of the liquid (the factors that determine the contact angle will be discussed in the following section). Although (2.10) was derived directly from the mechanical equilibrium condition which must exist across any curved interface, this is not the reason why the liquid rises in the capillary. This phenomenon occurs because the interfacial energy of the clean glass–water interface is much lower than that of the glass–air interface. The amount of energy released on wetting the glass surface and the potential energy gained by the liquid on rising in a gravitational field, must be minimized at equilibrium. Equation (2.10) can, in fact, be derived from this (free-energy minimization) approach, shown below. It is also interesting to note that because these interfacial energies are due to short-range forces, that is, surface properties, the capillary walls could be as thin as 100 Å and the liquid would still rise to exactly the same height (compare this with the gravitational force).

Capillary rise and the free energy analysis

The fundamental reason why a liquid will rise in a narrow capillary tube, against gravity, must be that $\gamma_{SV} > \gamma_{SL}$, i.e. that the free (surface) energy reduction on wetting the solid is balanced by the gain in gravitational potential energy. The liquid will rise to a height h, at which these factors are balanced. Thus, we must find the value of h for the equilibrium condition $dG_T/dh = 0$, where G_T is the total free energy of the system, at constant temperature. For a given height h:

$$\text{potential energy (increase)} = \pi r^2 hpg\frac{h}{2} \quad \text{(i.e. centre of gravity at } h/2)$$

$$\text{surface energy (decrease)} = 2\pi rh(\gamma_{SV} - \gamma_{SL})$$

$$\therefore \quad \frac{d\{\pi r^2 h^2 \rho g / 2 - 2\pi r h(\gamma_{SV} - \gamma_{SL})\}}{dh} = 0, \quad \text{at } h = h_{equil}$$

i.e.

$$\pi r^2 \rho g h - 2\pi r(\gamma_{SV} - \gamma_{SL}) = 0$$

$$\therefore \quad \gamma_{SV} - \gamma_{SL} = \frac{r\rho g h}{2}$$

and since

$$\gamma_{SV} = \gamma_{SL} + \gamma_{LV}\cos\theta \quad \text{(Young equation, see later)}$$

$$\therefore h = \frac{2\gamma_{LV}\cos\theta}{r\rho g}, \quad \text{as before}$$

The capillary rise method, although simple, is in practice, not as useful as the pendant drop method because of several experimental problems, such as the need to determine the contact angle, non-sphericity of the meniscus and uneven bore of the capillary.

One industrial application of the Laplace pressure generated in a pore is the use of Goretex membranes (porous Teflon membranes) to concentrate orange juice and other juices to reduce their bulk and hence transport costs. This process depends on the Laplace pressure retaining vapour in the Teflon pores, to allow water to be drawn through them as vapour, into a concentrated salt solution on the other side of the membrane. As can be seen from the simple calculation, see Figures 2.8 and 2.9, as long as the water contact angle remains high, say at around 110°, the pressure required to push water into the pores is greater than the hydrostatic pressure used in the operation and the juice can be successfully concentrated. Unfortunately, this process is very sensitive to the presence of surface-active ingredients in the juice, which can reduce the contact angle, allowing the pores to become filled with water and the juice become contaminated with salt. This process is illustrated in Figure 2.8. For this the Laplace pressures generated depend on the contact angle of water on the Teflon surface (Figure 2.9).

The dramatic effect of Laplace pressure can also be easily demonstrated using a syringe filled with water and attached to a Teflon

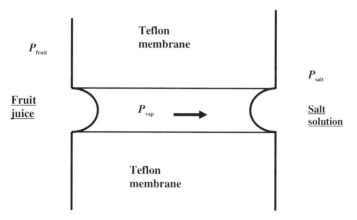

Figure 2.8 Schematic diagram of the concentration of fruit juice by vapour transport across a porous Teflon membrane.

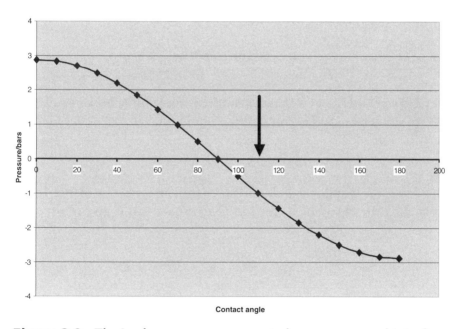

Figure 2.9 The Laplace pressure generated across a curved interface as a function of contract angle.

micron-sized membrane. Water cannot be pushed into the membrane; however, simply wetting the membrane with a droplet of ethanol will fill the pores and then the syringe easily pushes water through the membrane.

The Kelvin equation

It is often also important to consider the pressure of the vapour in equilibrium with a liquid. It can be demonstrated that this pressure, at a given temperature, actually depends on the curvature of the liquid interface. This follows from the basic equations of thermodynamics, given in Chapter 3, which lead to the result that

$$\partial\mu = +V_m \partial P$$

That is, the chemical potential of a component increases, linearly, with the total pressure of the system. (V_m is the partial molar volume of the component.) Thus, if we consider the change in chemical potential of the vapour and the liquid on producing a curved surface, we have the process shown in Figure 2.10. It follows that the change in chemical potential of the vapour is given by

$$\Delta\mu^v = \mu^v - \mu^0 = RT\ln\left(\frac{P}{P_0}\right)$$

Now, since both cases are at equilibrium, there must be an equivalent decrease in chemical potential of the liquid, that is,

$$\Delta\mu^v = \Delta\mu^l$$

But from the Laplace equation the change in pressure of the liquid (assuming the meniscus is, for simplicity, spherical) is given by

$$\Delta P = \frac{2\gamma}{r_e}$$

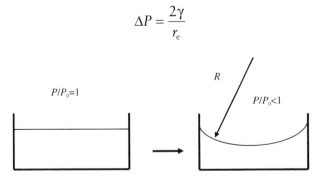

Figure 2.10 Schematic diagram showing that the equilibrium vapour pressure changes with the curvature of the liquid-vapour interface.

where r_e is the equilibrium radius of the (spherical) meniscus. Thus, it follows that the change in chemical potential of the liquid must be given by

$$\Delta\mu_l = V_m\Delta P = \frac{2\gamma V_m}{r_e}$$

which, combining with the earlier equation for the change in chemical potential of the vapour gives the result

$$\Delta\mu^{\nu} = \mu^{\nu} - \mu^0 = RT\ln\left(\frac{P}{P_0}\right) = \Delta\mu_l = \frac{2\gamma V_m}{r_e}$$

which on re-arrangement gives the Kelvin equation for spherical menisci:

$$RT\ln\left(\frac{P}{P_0}\right) = \frac{2\gamma V_m}{r_e} \quad \text{or, in general:} \quad RT\ln\left(\frac{P}{P_0}\right) = \gamma V_m\left(\frac{1}{R_1} + \frac{1}{R_2}\right)$$

This relationship gives some interesting and useful predictions for the behaviour of curved interfaces. For example, water at P/P_0 values of 0.99 should condense in cracks or capillaries and produce menisci of (negative) radius 105 nm, of the type shown in Figure 2.11. However, for a sessile droplet, there must be a positive Kelvin radius, and for typical large droplets of, say, mm radius they must be in equilibrium with vapour very close to saturation (Figure 2.12). A range of calculated values for water menisci at 21°C are given in Figure 2.13 for both concave (negative-radius) and convex (positive-radius) menisci.

Another common method used to measure the surface tension of liquids is called the 'Wilhelmy plate'. These methods use the force (or

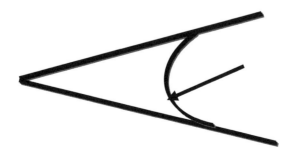

Figure 2.11 Capillary condensation of water vapour into a crack.

Figure 2.12   Diagram of a sessile droplet.

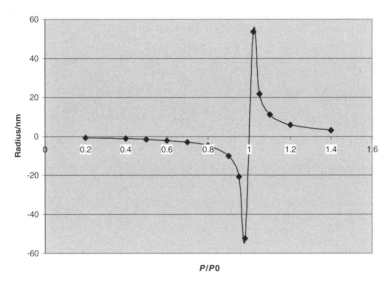

Figure 2.13   Graph of relative vapour pressure against radius of the corresponding equilibrium meniscus.

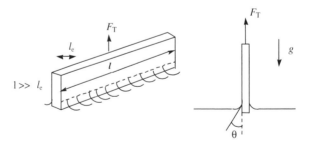

Figure 2.14   Diagram of the Wilhelmy plate method for measuring the surface tension of liquids.

tension) associated with a meniscus surface to measure the surface energy rather than using the Laplace pressure equation. (Note that in real cases both factors usually arise but often only one is needed to obtain a value for γ.) The Wilhelmy plate is illustrated in Figure 2.14. The total force F_T (measured using a balance) is given by

$$F_\text{T} = F_\text{W} + 2l\gamma\cos\theta \qquad (2.11)$$

where F_W is the dry weight of the plate. (Note that the base of the plate is at the same level as the liquid thereby removing any buoyancy forces.) The plates are normally made of thin platinum which can be easily cleaned in a flame and for which l_e can be ignored. Again, this method has the problem that θ must be known if it is greater than zero. In the related du Noüy ring method, the plate is replaced by an open metal wire ring. At the end of this chapter, a laboratory class is used to demonstrate yet another method, which does not require knowledge of the contact angle and involves withdrawal of a solid cylinder attached to a liquid surface.

The surface energy and cohesion of solids

Measurement of the surface energy of a liquid is relatively easy to both perform and understand. All methods are based on measuring the work required to create a new surface by transferring molecules from bulk liquid. However, what about the surface energy of a solid? Clearly, for solids it is impractical to move molecules from bulk to the surface. There are basically two ways by which we can attempt to obtain the surface energy of solids:

1. by measuring the cohesion of the solid, and

2. by studying the wetting behaviour of a range of liquids with different surface tensions on the solid surface.

Neither methods is straightforward and the results are not as clear as those obtained for liquids. The cohesive energy per unit area, W_c, is equal to the work required to separate a solid in the ideal process illustrated in Figure 2.15. In this ideal process the work of cohesion, W_c, must be equal to twice the surface energy of the solid, γ_s. Although this appears simple as a thought experiment, in practice it is difficult. For example, we might measure the critical force (F_c) required to separate the material but then we need a theory to relate this to the total work done. The molecules near the surface of the freshly cleaved solid will rearrange after measuring F_c. Also, the new area will not usually be smooth and hence the true area is much larger than the geometric area.

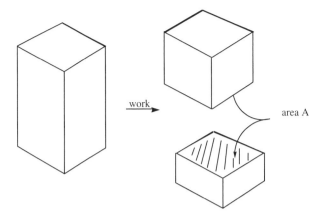

Figure 2.15 Ideal experiment to measure the work required to create new area and hence find the surface energy of a solid.

Only a few materials can be successfully studied in this way. One of them is the layered natural aluminosilicate crystal, muscovite mica, which is available in large crystals and can be cleaved in a controlled manner to produce two molecularly smooth new surfaces.

In comparison, the adhesive energy per unit area W_a between two different solids is given by:

$$W_a = \gamma_A + \gamma_B - \gamma_{AB} \qquad (2.12)$$

where γ_A and γ_B are the surface energies of the solids and γ_{AB} is the interfacial energy of the two solids in contact ($\gamma_{AA} = 0$). Again the adhesive energy is a difficult property to measure. It is also very hard to find the actual contact area between two different materials since this is almost always much less than the geometric area. That this is the case is the reason why simply pressing two solids together does not produce adhesion (except for molecularly smooth crystals like mica) and a 'glue' must be used to dramatically increase the contact area. The main function of a glue is to facilitate intimate molecular contact between two solids, so that strong short-range van der Waals forces can hold the materials together.

The contact angle

The second approach to obtaining the surface energies of solids involves the study of wetting and non-wetting liquids on a smooth, clean solid substrate. Let us examine the situation for a non-wetting

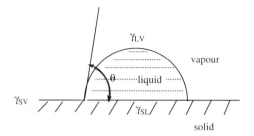

Figure 2.16 Diagram of a sessile droplet.

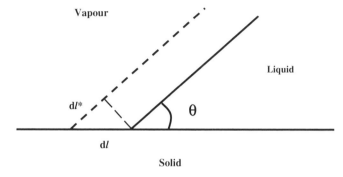

Figure 2.17 Diagram of the three phase line and its perturbation to determine the contact angle.

liquid (where $\theta > 0°$), which will form a sessile drop on the surface of a solid (Figure 2.16). Using an optical microscope, it is possible to observe and measure a finite contact angle (θ) as the liquid interface approaches the three-phase-contact perimeter of the drop. Let us consider the local equilibrium situation along a small length of the 'three-phase line' or TPL. This is the line where all three phases are in contact. Let us examine this region in more detail in the schematic diagram, Figure 2.17. Let us examine the equilibrium contact angle, θ, for which an infinitesimal movement in the TPL by distance dl to the left-hand side, will not change the total surface free energy of the system. We can consider area changes for each of the three interfaces for unit length 'l' vertical to the page and along the TPL. Thus, the total interfacial energy change must be given by the sum

$$dG = \gamma_{sl} l dl + \gamma_{lv} l dl^* - \gamma_{sv} l dl$$

From simple geometry, $dl^* = dl \cos\theta$ and hence at equilibrium, where $dG/dl = 0$, it follows that

$$\gamma_{SV} = \gamma_{SL} + \gamma_{LV}\cos\theta \qquad (2.13)$$

This important result is called the *Young equation*. It can also be derived by simply considering the horizontal resolution of the three surface tensions (i.e. as forces per unit distance), via standard vector addition (Figure 2.18). However, what becomes of the vertical component? This force is actually balanced by the stresses in the solid around the drop perimeter (or TPL), which can actually be visually observed on a deformable substrate, such as paraffin wax.

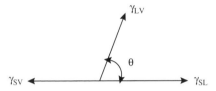

Figure 2.18 Balance of energies at the TPL gives the Young equation directly.

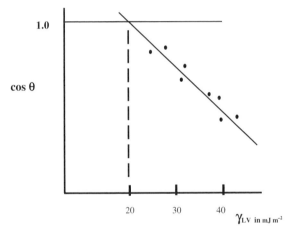

Figure 2.19 Typical plot of the contact angles of a range of liquids on a low energy solid.

Since we can measure the liquid surface energy, γ_{LV}, the value of (γ_{SV} $- \gamma_{SL}$) can be obtained, but, unfortunately, γ_{SL} is as difficult to measure directly as γ_{SV}. However, if θ is measured for a range of liquids with different surface energies, then a plot of $\cos\theta$ against γ_{LV} gives a 'critical surface energy' value, γ_c, at $\theta = 0°$ (the complete wetting case). It is often not unreasonable to equate γ_c with γ_{SV} because in many cases at complete wetting γ_{SL} approaches zero. The schematic Figure 2.19 corresponds to the type of behaviour observed for a range of different liquids wetting Teflon. The low surface energy of Teflon has been estimated from this type of data.

Clearly the surface energy of a solid is closely related to its cohesive strength. The higher the surface energy, the higher its cohesion. This has some obvious and very important ramifications. For example, the strength of a covalently bonded solid, such as a glass or metal, must always be greatest in a high vacuum, where creation of new surface must require the greatest work. The strength of the same material in water vapour or immersed in liquid water will be much reduced, often by at least an order of magnitude. This is because the freshly formed solid surface must initially be composed of high-energy atoms and molecules produced by the cleavage of many chemical bonds. These new high-energy surfaces will rapidly adsorb and react with any impingent gas molecules. Many construction materials under strain will therefore behave differently, depending on the environment. It should also be noted that the scoring of a glass rod only goes to a depth of about 0.01 per cent of the rod's thickness but this still substantially reduces its strength. Clearly, crack propagation determines the ultimate strength of any material and, in general, cracks will propagate more easily in an adsorbing environment (e.g. of liquid or vapour). Objects in outer space can, therefore, be produced using thinner materials but still with the same strength.

A list of (advancing) water contact angles on various solid substrates is given in Table 2.1. It is immediately obvious that water will not wet 'low-energy' surfaces ($\gamma_{SV} < 70\,\text{mJ}\,\text{m}^{-2}$) such as hydrocarbons, where there is no possibility of either hydrogen bonding or dipole–dipole interactions with the solid substrate. However, complete wetting occurs

Table 2.1

	θ_A
Paraffin wax	110°
PTFE (teflon)	108°
Polyethylene	95°
Graphite	86°
AgI	70°
Mica	7°
Gold (clean)	0°
Glass (clean)	0°
(Hg on glass	135°)

on 'high-energy' surfaces ($\gamma_{SV} > 70\,\mathrm{mJ\,m^{-2}}$), such as clean glass and most metals. Directly from our concept of surface energies, it is clear that we would expect a liquid to spread on a substrate if:

$$\gamma_{SV} - \gamma_{LV} - \gamma_{SL} > 0 \qquad (2.14)$$

In fact, we can define a parameter

$$S_{LS} = \gamma_{SV} - \gamma_{LV} - \gamma_{SL} \qquad (2.15)$$

called the *spreading coefficient*. If $S_{LS} > 0$, the liquid will spread. Since γ_{LV} for mercury is about $485\,\mathrm{mJ\,m^{-2}}$ (because of very strong metallic bonding), we would not expect this liquid to spread on anything but very high-energy surfaces (such as other metals).

The contact angle is usually quite easy to measure and is a very useful indicator for the wetting properties of a material. Whether water will wet a solid or not is used extensively by the mining industry in a separation technique called 'froth flotation'. As mentioned previously, a rock containing a required mineral together with unwanted solid, often quartz, can be ground to give colloidal-size particles of the pure mineral. If the powder is dispersed in water and bubbles are continuously passed through the chamber, those particles with water contact angles greater than about 20° will attach to the rising bubbles, whereas the 'wetting' particles will remain dispersed (Figure 2.20).

In water the wetted solid is termed 'hydrophilic', whereas the non-wetted solid is 'hydrophobic'. Naturally hydrophobic minerals, such as some types of coal, talc and molybdenite are easily separated from the unwanted hydrophilic quartz sand (referred to as 'gangue'). However, surfactants and oils are usually added as 'collectors'. These compounds adsorb onto the hydrophilic mineral surface and make it hydrophobic.

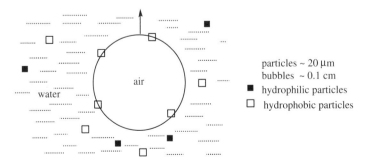

Figure 2.20 Diagram of a bubble 'collecting' hydrophobic particles in a flotation process.

Frothers are also added to stabilize the foam at the top of the chamber, so that the enriched mineral can be continuously scooped off. The selective flotation of a required mineral depends critically on surface properties and these can be carefully controlled using a wide range of additives. Throughout the world a large quantity (about 10^9 tons annually) of minerals are separated by this method.

Although the contact angle is a very useful indicator of the energy of a surface, it is also affected by a substrate's surface roughness and chemical micro-heterogeneity. This can be well illustrated by comparing ideal, calculated contact angle values with measured ones. For example, we can easily calculate the expected water contact angle on a liquid or solid pure hydrocarbon surface by using the surface tension of water and hexadecane ($27.5\,\mathrm{mN\,m^{-1}}$) and the interfacial tension between the oil and water ($53.8\,\mathrm{mN\,m^{-1}}$). Use of these values in the Young equation gives an expected angle of 111°, in close agreement with the values observed for paraffin wax. However, on many real surfaces the observed angle is hysteretic, giving quite different values depending on whether the liquid droplet size is increasing (giving the advancing angle θ_A) or decreasing (giving the receding angle θ_R). The angles can differ by as much as 60° and there is some controversy as to which angle should be used (for example in the Young equation) or even if the average value should be taken. Both θ_A and θ_R must be measured carefully, whilst the three-phase line is stationary but just on the point of moving, either forwards or back. In general, both angles and the differences between the two give indirect information about the state of the surface, and both should always be reported. The degree of hysteresis observed is a measure of both surface roughness and surface chemical heterogeneity.

Industrial Report

Photographic-quality printing

Modern photographic-quality inkjet papers have a surface coating comprising either a thin polymer film or a fine porous layer. In either case the material is formed using a high-speed coating process. This process requires careful control to obtain the necessary uniformity together with

Continued

low manufacturing costs. From a scientific point of view, all coating processes have at least one static wetting line and one dynamic wetting line, and in many cases the behaviour of both the dynamic and static wetting lines determines the speed and uniformity of the final coating. At a fundamental level, surface forces control wetting line behaviour. However, the details of the physical and chemical behaviour of a dynamic wetting line are still a matter of scientific debate. Having created a thin-film coating, the liquid sheet is susceptible to a variety of potential defects ranging from holes generated through Marangoni-driven flows to mottle driven by interface instabilities. These flow defects can, and are, controlled by the use of appropriate surfactant additives.

Porous inkjet papers are in general created from colloidal dispersions. The eventual random packing of the colloid particles in the coated and dried film creates an open porous structure. It is this open structure that gives photographic-quality inkjet paper its 'apparently dry' quality as it comes off the printer. Both the pore structure and pore wettability control the liquid invasion of the coated layer and therefore the final destination of dyes. Dispersion and stability of the colloidal system may require dispersant chemistries specific to the particle and solution composition. In many colloidal systems particle–particle interactions lead to flocculation which in turn leads to an increase in viscosity of the system. The viscosity directly influences the coating process, through the inverse relation between viscosity and maximum coating speed.

Surface and colloid science can also play a significant role in formulation of pigmented inks, another colloidal dispersion again requiring a good dispersant for stability within the ink cartridge. Jettability of the ink from the printhead and the wetting behaviour of the ink on and in the paper are both controlled by surface interactions. Inkjet material manufacture and design therefore provides a fertile ground for the surface and colloid scientist to apply their skills.

Dr Andrew Clarke
Surface and Colloid Science Group
Research & Development, Kodak Limited
Harrow, UK

Sample problems

1. What pressure must be applied to force water through an initially dry Teflon membrane which has a uniform pore size of $0.5\,\mu m$ diameter? What factors can reduce this pressure? Give examples of the industrial and everyday use of this type of porous material. ($\theta_{water} = 110°$, $\gamma_{water} = 73\,mN\,m^{-1}$.)

2. Use the Laplace equation to calculate the spherical radius of the soap film which is formed by the contact of two bubbles with radii of 1 and 3 cm. Assume that the soap bubbles have a surface tension of $30\,mJ\,m^{-2}$. Draw a sketch of the contacting bubbles to help you.

3. Calculate the surface tension of a wetting liquid, of density 1.2 g/ml, which rises 1.05 m in a vertical capillary tube, with an internal diameter of 0.2 mm.

4. Two curved regions on the surface of a water droplet have principal radii of 0.2, 0.67 and 0.1, 0.5 cm. What is their difference in vertical height?

5. Water can just be forced through a Goretex (porous Teflon) membrane at an applied pressure of 1.5 bar. What is the pore size of the mebrane?

6. Explain why it is possible to make sandcastles only in partially wet sand and not in either fully immersed or completely dry sand.

7. Why do construction workers spray water when they are working with very dusty materials?

8. Figure 2.21 shows two spherical, hydrophilic solid particles held together by a water meniscus. If the upper particle is held fixed, calculate the minimum force (F) required to pull the lower particle away. Assume that water has a zero contact angle with the solid.

9. A colloidal particle is held on to a solid surface by a water meniscus, as illustrated in Figure 2.22. Estimate the minimum force (F) required to detach the particle. (Ignore the mass of the colloid.)

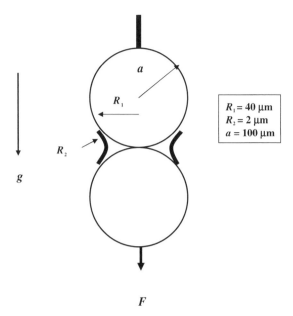

Figure 2.21 Two spherical particles held together by a water meniscus.

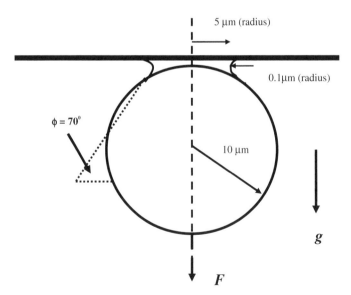

Figure 2.22 Colloidal particle attached to a solid surface by a water meniscus.

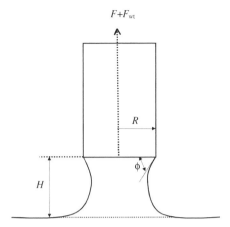

Figure 2.23   Schematic diagram of a cylinderical rod pulled from a liquid surface.

Experiment 2.1   Rod in free surface (RIFS) method for the measurement of the surface tension of liquids

Introduction

There is a wide range of different methods used to measure surface tension/energy of liquids and solutions. The RIFS technique has the advantage that it requires only simple, easily available (cheap) equipment and yet gives absolute and accurate surface energies (to $0.1\,\mathrm{mJ\,m^{-2}}$). The principle of this method is that the maximum force on a cylindrical rod pulled through the surface of a liquid (as shown in Figure 2.23) is related to the surface tension of the liquid.

The rod lifts up a meniscus above the surface of the liquid and the weight of this meniscus can be measured. The additional force on the rod (i.e. above its own weight) must be given by

$$F = \pi R^2 H \rho g + 2\pi R \gamma \sin \Phi \qquad (2.16)$$

The first term on the right-hand side is due to the hydrostatic pressure (or suction) acting over the base of the rod and the second term is due to surface tension forces around the perimeter. (It should be noted that Φ is not equal to the equilibrium contact angle θ but is determined by the meniscus shape.)

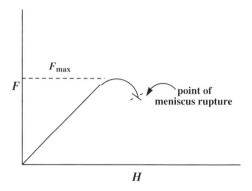

Figure 2.24 Illustration of the force on the rod as a function of height above the liquid surface.

If we could accurately photograph the profile illustrated in Figure 2.23 we could measure H, R and Φ and since F is simply the excess weight on the rod we could obtain γ. However, experimentally this would be difficult, and significant errors are likely in measuring the height H.

A much easier method has been developed by Padday et al. (1975) (*Faraday Trans. I*, 71, 1919), which only requires measurement of the maximum force or weight on the rod as it is pulled upwards. It has been shown by using the Laplace equation to generate meniscus profiles that this maximum is stable and quite separate from the critical pull-off force where the meniscus ruptures. A typical force–height curve is shown in Figure 2.24.

Padday et al. derived an equation which allows the surface tension of the liquid to be obtained from the measured maximum force on the rod, the known density of the liquid (ρ) and the radius of the cylinder. This equation is

$$R/k = a_0 + a_1\left(R^3/V\right) + a_2\left(R^3/V\right)^2 + a_3\left(R^3/V\right)^3 \tag{2.17}$$

where $k = (\gamma/\rho g)^{0.5}$ and V is the volume of the meniscus raised above the level of the liquid, which is directly obtained from the measured weight of the meniscus. a_0, a_1, a_2 and a_3 are constants which depend on the value of R^3/V and are given in Table 2.2. Once the value of R^3/V is measured the correct values of a_n are chosen and used to calculate k and hence γ.

Table 2.2

R^3/V	a_0	a_1	a_2	a_3
0.01–0.02	9.07578×10^{-2}	2.07380×10^{1}	-4.46445×10^{2}	6.23543×10^{3}
0.02–0.03	1.15108×10^{-1}	1.64345×10^{1}	-2.00113×10^{2}	1.63165×10^{3}
0.03–0.04	1.06273×10^{-1}	1.69246×10^{1}	-2.04837×10^{2}	1.57343×10^{3}
0.04–0.05	6.34298×10^{-2}	1.88348×10^{1}	-2.20374×10^{2}	1.43729×10^{3}
0.05–0.07	1.56342×10^{-1}	1.23019×10^{1}	-6.96970×10^{1}	2.93803×10^{2}
0.07–0.10	2.21619×10^{-1}	9.31363	-2.39480×10^{1}	5.96204×10^{1}
0.10–0.15	3.11064×10^{-1}	6.97932	-3.58929	0.0
0.15–0.20	3.67250×10^{-1}	6.26621	-1.32143	0.0
0.20–0.30	4.40580×10^{-1}	5.60569	1.63171×10^{-1}	0.0
0.30–0.40	4.47385×10^{-1}	5.63077	0.0	0.0
0.40–0.50	4.72505×10^{-1}	5.39906	4.24569×10^{-1}	0.0
0.50–0.60	3.780×10^{-1}	5.80000	0.0	0.0
0.60–0.80	5.72110×10^{-1}	5.15631	5.33894×10^{-1}	0.0
0.80–1.00	2.99048×10^{-1}	5.86260	7.83455×10^{-2}	0.0
1.00–1.20	6.76415×10^{-1}	5.16281	4.01204×10^{-1}	0.0
1.20–1.40	4.08687×10^{-2}	6.20312	-2.40752×10^{-2}	0.0
1.40–1.60	2.53174×10^{-1}	5.90351	8.14259×10^{-2}	0.0
1.60–1.85	-1.30000×10^{-2}	6.20000	0.0	0.0

Experimental details

This experiment requires skill and careful technique in order to obtain accurate results. As with most surface chemistry experiments cleanliness is of paramount importance. High-energy liquids such as water easily pick up surface-active contaminants from the air in a laboratory and great care should be taken to reduce exposure. Contaminants generally do not adsorb at the surface of low-energy liquids such as hexane and hence are less of a problem.

The liquids to be studied in this experiment are water, hexane, *n*-octanol and aqueous solutions of CTAB. It is recommended that they be measured in the order written, where the most critical with respect to contamination is first. The water used should be the best available, such as double distilled, and should be stored in a sealed flask before use. Pure samples of the other liquids should also be used as well as top-quality water to make up the CTAB solutions. The CTAB solutions should be measured at concentrations of 0.01, 0.1, 0.3, 0.6, 1 and 10 mM at a temperature above 21°C. CTAB has a Krafft temperature around 20°C – below this temperature the surfactant will precipitate from aqueous solution at the higher concentrations (see later).

The apparatus used to measure the maximum force on a rod is shown in Figure 2.25. Be extremely careful in connecting the rod to the balance hook (the balance should be switched off during this procedure). An electronic balance is usually more robust and easier to use. The mechanical stage is used to roughly position the level of the liquid, and the level is then gradually and carefully lowered by withdrawing liquid into the motor-driven syringe. Assuming that there are no vibrations and the rod is almost perfectly vertical it should be possible to raise the rod beyond the force maximum without rupturing the meniscus. The maximum weight can then be accurately found by adding and removing liquid in this region. (This must be done very slowly for the CTAB solutions to allow equilibrium adsorption of the surfactant.) The measurement is acceptable only when the meniscus is symmetrical around the perimeter of the base of the rod. The temperature of the liquid should be noted. The maximum weight of the meniscus is equal to the maximum measured weight minus the dry weight of the rod.

The syringe, dish and rod must be thoroughly cleaned on changing liquids. For the water measurements Pyrex glass and the stainless steel rod are best cleaned by (analar) ethanol scrubbing and washing followed by high-quality water rinsing. For the organic liquids ethanol rinsing is sufficient with final rinsing using the liquid to be measured. (Note that the rod must be free of excess liquid during measurement and also there must be no liquid film on the sides of the rod.)

The base diameter of the rod must be accurately measured using a micrometer and taking an average over at least three positions around the base. That this measurement is critical can be seen from an examination of Equation (2.17). The perimeter of the base must be sharp and undamaged.

Calculate the surface energies of each of these liquids and plot a graph of γ for the CTAB solutions as a function of $\log_{10}(\text{conc.})$. Use your results and the Gibbs adsorption equation (see later) to estimate the minimum surface area per CTAB molecule adsorbed at the air–water interface.

Questions

1. What advantages does this (RIFS) technique have over the Wilhelmy plate method?

2. Why is the surface energy of octanol higher than that of hexane?

3. What is the cmc of CTAB solutions?

4. Why do we plot the CTAB concentration as a log function?

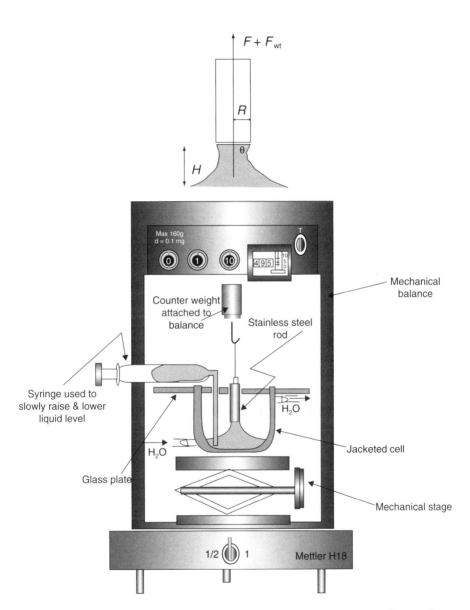

Figure 2.25   Diagram of the apparatus used to measure the surface tension of liquids by the RIFS method.

Experiment 2.2 Contact angle measurements

Introduction

The value of the equilibrium contact angle (θ) at the three-phase line (TPL) produced by a liquid droplet placed on a flat, solid substrate is determined by the balance of interfacial energies at each surface. Thomas Young derived an equation describing this situation in 1804:

$$\gamma_{SV} = \gamma_{SL} + \gamma_{LV} \cos\theta \qquad (2.18)$$

where γ_{SV}, γ_{SL} and γ_{LV} are the energies of the solid/vapour, solid/liquid and liquid/vapour interfaces. The contact angle θ is the angle subtended between the liquid/vapour and liquid/solid interfaces and is illustrated in the following Figure 2.26. In order for the droplet to be at equilibrium the gas phase must be saturated with the liquid vapour. For this reason measurements are normally carried out in a sealed vessel with good temperature control.

Contact angle measurements are of fundamental importance in a range of industrial and everyday processes such as flotation, painting (i.e. the paint must wet the substrate) and weather-proofing. In the flotation process a solid block of the powdered mineral to be floated is often studied using a wide range of collector (i.e. surfactant) solutions to determine optimum flotation conditions.

The surface energies of liquids can be directly measured, but this is not the case for most solid/vapour and solid/liquid energies (with one or two notable exceptions, such as mica). An estimate of solid/vapour energies can, however, be obtained by measuring the equilibrium contact angles of a range of different liquids of higher surface energy

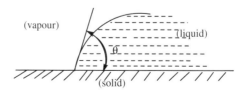

Figure 2.26 Contact angle made by a sessile droplet.

than the substrate, i.e. non-wetting liquids. A Zisman plot of $\cos\theta$ against γ_{LV} is often linear and can be extrapolated to $\cos\theta = 1$ (i.e. $\theta = 0°$) to give the value of the 'critical surface tension' γ_c of the solid, where a liquid will just completely wet the surface. It has been shown that γ_c is a reasonable approximation for γ_{SV}.

In this experiment γ_c will be determined for methylated (hydrophobic) soda glass. Both advancing (θ_A) and receding (θ_R) angles will be measured in order to estimate the degree of hysteresis in each case. Use the values of both θ_A and θ_R, as well as the average, to produce several Zisman plots.

Experimental details

Contact angle measurement

The contact angles can be measured by observing the TPL through a microscope that has a rotating eyepiece and cross-hairs. The eyepiece is used to measure the angle of the cross-hairs on a protractor scale. By tilting the microscope slightly the reflected image of the liquid droplet can also be observed and the double-angle (2θ) can be measured, which increases the accuracy. An illustration of this type of apparatus is given in Figure 2.27.

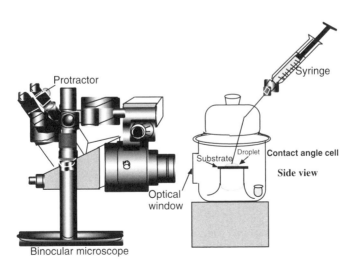

Figure 2.27   Schematic diagram of the contact angle apparatus.

The glass cell and syringe should be cleaned by ethanol rinsing followed by drying using nitrogen. The glass plate (which must be cleaned in the manner described later) is then positioned on the upraised table and a small beaker of the liquid to be used is placed in the base of the cell. The cell is then sealed and left to equilibrate for 5 min. About 0.5–1.0 cm^3 of the liquid is filled into the syringe (taking care to remove all air bubbles) and the steel needle is lowered to within 1 mm of roughly the centre of the plate (use the microscope for this). A droplet of the liquid is then slowly forced out and allowed to equilibrate. Since the needle is left in the droplet during measurement, the droplet must be of sufficient size that the region near the TPL line is not affected by the needle changing the shape of the droplet. The drop volume should then be slowly increased until the maximum angle is obtained just before the TPL moves forwards. (This has to be repeated several times to obtain the maximum angle.) Measure the contact angle with the TPL at different positions on the plate. Take the average value to give the advancing contact angle (θ_A).

Follow a similar procedure to measure the receding angle (θ_R) but slowly withdraw liquid into the syringe and measure the minimum angle just before the TPL moves.

Sample preparation

Before methylation, the soda glass plates must be cleaned by washing with warm 10% NaOH solution and then rinsed in high-quality distilled water, finally blow-drying in nitrogen. Be careful and always wear safety glasses. Warm, concentrated NaOH solution is very harmful to the eyes. The contact angle of water on clean glass should be very low; otherwise, further cleaning is required. The cleaned glass is very easily contaminated by finger grease and exposure to laboratory air. For these reasons the samples have to be prepared just before they are used in the cell and must only be handled using clean tweezers.

Methylated (hydrophobic) glass is prepared simply by exposing a clean, dry plate to the vapour of highly reactive trimethylchlorosilane (Me_3SiCl) for about 1 minute in a fume cupboard. Simply place the plate with the polished surface exposed in a large, clean beaker containing a smaller beaker of liquid Me_3SiCl. Loosely cover the large beaker. The Me_3SiCl reacts vigorously with water as well as the surface silanol groups on glass (Figure 2.28).

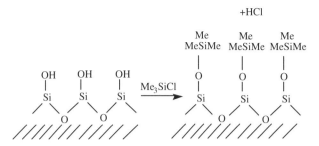

Figure 2.28   Methylation of the silica surface.

Table 2.3

Liquid	$\gamma_{LV}(mJ\,m^{-2})$
water	73
formamide	56
methylene iodide	49
(or ethylene glycol	47.5)
propylene carbonate	41
di-methyl aniline	36.5

The hydrochloric acid vapour evolved is toxic and all reactions should be carried out in a fume cupboard.

The resulting monolayer of methyl groups is chemically attached and completely alters the wetting properties of the surface. Again, to prevent contamination these plates must be handled only with tweezers and stored in cleaned, sealed containers.

Liquids used to determine the critical surface tension
of methylated glass

Measure advancing and receding contact angles on methylated glass plates using the liquids in the order given in Table 2.3.

For all these liquids except water prepare the syringe and cell in a fume cupboard and seal before measuring contact angles in the laboratory (use caution). Use the same methylated plate for the first two liquids and a second plate for the other three. In each case rinse the plate with clean ethanol and blow dry when changing liquids.

Plot out $\cos\theta_A$, $\cos\theta_R$ and $\cos\theta_{AV}$ against the corresponding γ_{LV} values to estimate the value of γ_c.

At the end of the experiment clean the cell and syringe thoroughly using ethanol and blow dry with nitrogen.

Questions

1. Why is it that only a very thin monolayer (~5 Å) of adsorbed methyl groups completely alters the macroscopic water-wetting properties of glass?

2. What do you think are the causes of the contact angle hysteresis observed in this experiment?

3. What would you expect for the wetting properties of these liquids on untreated, clean glass?

4. Is the value you obtained for γ_c reasonable for this type of surface?

3

Thermodynamics of Adsorption

Derivation of the Gibbs adsorption isotherm. Determination of the adsorption of surfactants at liquid interfaces. Laboratory project to determine the surface area of the common adsorbent, powdered activated charcoal.

Basic surface thermodynamics

For an open system of variable surface area, the Gibbs free energy must depend on composition, temperature, T, pressure, p, and the total surface area, A:

$$G = G(T, p, A, n_1, n_2, \ldots, n_k) \tag{3.1}$$

From this function it follows that:

$$dG = \left(\frac{\partial G}{\partial T}\right)_{p,n_i} dT + \left(\frac{\partial G}{\partial p}\right)_{T,n_i} dp$$

$$+ \left(\frac{\partial G}{\partial A}\right)_{T,p,n_i} dA + \sum_{i=1}^{i=k} \left(\frac{\partial G}{\partial n_i}\right)_{T,p,n_j} dn_i \tag{3.2}$$

Applied Colloid and Surface Chemistry Richard M. Pashley and Marilyn E. Karaman
© 2004 John Wiley & Sons, Ltd. ISBN 0 470 86882 1 (HB) 0 470 86883 X (PB)

The first two partial differentials refer to constant composition, so we may use the general definitions

$$G = H - TS = U + PV - TS \tag{3.3}$$

to obtain

$$\left(\frac{\partial G}{\partial T}\right)_{p,n_i} = -S \tag{3.4}$$

and

$$\left(\frac{\partial G}{\partial p}\right)_{T,n_i} = V \tag{3.5}$$

Insertion of these relations into (3.2) gives us the fundamental result

$$dG = -SdT + Vdp + \gamma dA + \sum_{i=1}^{i=k} \mu_i dn_i \tag{3.6}$$

where the chemical potential μ_i is defined as

$$\mu_i \equiv \left(\frac{\partial G}{\partial n_i}\right)_{T,p,n_j} \tag{3.7}$$

and the surface energy γ as

$$\gamma = \left(\frac{\partial G}{\partial A}\right)_{T,p,n_i} \tag{3.8}$$

The chemical potential is defined as the increase in free energy of a system on adding an infinitesimal amount of a component (per unit number of molecules of that component added) when T, p and the composition of all other components are held constant. Clearly, from this definition, if a component 'i' in phase A has a higher chemical potential than in phase B (that is, $\mu_i^A > \mu_i^B$) then the total free energy will be lowered if molecules are transferred from phase A to B and this will occur in a spontaneous process until the chemical potentials equalize, at equilibrium. It is easy to see from this why the chemical potential is so useful in mixtures and solutions in matter transfer (open) processes. This is especially clear when it is understood that μ_i is a simple function of concentration, that is:

$$\mu_i = \mu_i^\circ + kT \ln C_i. \tag{3.9}$$

for dilute mixtures, where μ_i° is the standard chemical potential of component 'i', usually 1 M for solutes and 1 atm for gas mixtures. This equation is based on the entropy associated with a component in a mixture and is at the heart of why we generally plot measurable changes in any particular solution property against the log of the solute concentration, rather than using a linear scale. Generally, only substantial changes in concentration or pressure produce significant changes in the properties of the mixture. (For example, consider the use of the pH scale.)

Derivation of the Gibbs adsorption isotherm

Let us consider the interface between two phases, say between a liquid and a vapour, where a solute (i) is dissolved in the liquid phase. The real concentration gradient of solute near the interface may look like Figure 3.1. When the solute increases in concentration near the surface (e.g. a surfactant) there must be a surface excess of solute n_i^σ, compared with the bulk value continued right up to the interface. We can define a surface excess concentration (in units of moles per unit area) as:

$$\Gamma_i = \frac{n_i^\sigma}{A} \tag{3.10}$$

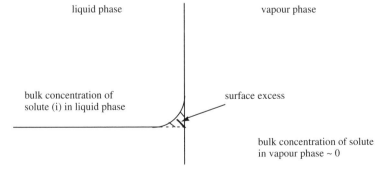

Figure 3.1 Diagram of the variation in solute concentration at an interface between two phases.

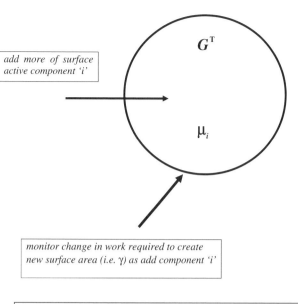

add more of surface
active component 'i'

G^T

μ_i

monitor change in work required to create
new surface area (i.e. γ) as add component 'i'

**Note that the total free energy of the system always
increases on addition of component 'i' and when its
surface area is increased.**

Figure 3.2 Diagramatic illustration of the change in surface energy
caused by the addition of a solute.

where A is the interfacial area (note that Γ_i may be either positive or
negative).

Let us now examine the effect of adsorption on the interfacial energy
(γ). If a solute 'i' is positively adsorbed with a surface density of Γ_i, we
would expect the surface energy to *decrease* on increasing the bulk concentration of this component (and vice versa). This situation is illustrated in Figure 3.2, where the total free energy of the system G^T and
μ_i are both increased by addition of component i but because this
component is favourably adsorbed at the surface (only relative to the
solvent, since both have a higher energy state at the surface), the work
required to create new surface (i.e. γ) is reduced. Thus, although
the total free energy of the system increases with the creation of new
surface, this process is made easier as the chemical potential of the
selectively adsorbed component increases (i.e. with concentration). This
reduction in surface energy must be directly related to the change in
chemical potential of the solute and to the amount adsorbed and is
therefore given by the simple relationship:

$$d\gamma = -\Gamma_i d\mu_i \qquad (3.11)$$

or, for the case of several components,

$$d\gamma = -\Sigma_i \Gamma_i d\mu_i \qquad (3.12)$$

The change in μ_i is caused by the change in bulk solute concentration. This is the Gibbs surface tension equation. Basically, these equations describe the fact that increasing the chemical potential of the adsorbing species reduces the energy required to produce new surface (i.e. γ). This, of course, is the principal action of surfactants, which will be discussed in more detail in a later section.

Using this result let us now consider a solution of two components

$$d\gamma = -\Gamma_1 d\mu_1 - \Gamma_2 d\mu_2 \qquad (3.13)$$

and hence the adsorption excess for one of the components is given by

$$\Gamma_1 = -\left(\frac{\partial\gamma}{\partial\mu_1}\right)_{T,\mu_2} \qquad (3.14)$$

Thus, in principle, we could determine the adsorption excess of one of the components from surface tension measurements, if we could vary μ_1 independently of μ_2. But the latter appears not to be possible, because the chemical potentials are dependent on the concentration of each component. However, for dilute solutions the change in μ for the solvent is negligible compared with that of the solute. Hence, the change for the solvent can be ignored and we obtain the simple result that

$$d\gamma = -\Gamma_1 d\mu_1 \qquad (3.15)$$

Now, since $\mu_1 = \mu_1^0 + RT\ln c_1$, differentiation with respect to c_1 gives

$$\left(\frac{\partial\mu_1}{\partial c_1}\right)_T = RT\left(\frac{\partial\ln c_1}{\partial c_1}\right)_T = \frac{RT}{c_1} \qquad (3.16)$$

Then substitution in (3.15) leads to the result:

$$\Gamma_1 = -\frac{1}{RT}\left(\frac{\partial \gamma}{\partial \ln c_1}\right)_T = \frac{c_1}{RT}\left(\frac{\partial \gamma}{\partial c_1}\right)_T \qquad (3.17)$$

This is the important Gibbs adsorption isotherm. (Note that for concentrated solutions the activity should be used in this equation.) Experimental measurements of γ over a range of concentrations allows us to plot γ against $\ln c_1$ and hence obtain Γ_1, the adsorption density at the surface. The validity of this fundamental equation of adsorption has been proven by comparison with direct adsorption measurements. The method is best applied to liquid/vapour and liquid/liquid interfaces, where surface energies can easily be measured. However, care must be taken to allow equilibrium adsorption of the solute (which may be slow) during measurement.

Finally, it should be noted that (3.17) was derived for the case of a single adsorbing solute (e.g. a non-ionic surfactant). However, for ionic surfactants such as CTAB, two species (CTA^+ and Br^-) adsorb at the interface. In this case the equation becomes

$$\Gamma_1 = -\frac{1}{2RT}\left(\frac{\partial \gamma}{\partial \ln c_1}\right)_T \qquad (3.18)$$

because the bulk chemical potentials of both ions change (equally) with concentration of the surfactant.

Question: Consider which form of the isotherm would apply to an ionic surfactant solution made up in an excess of electrolyte.

Determination of surfactant adsorption densities

Typical results obtained for the variation in surface tension with surfactant (log) concentration, for a monovalent surfactant, are given in Figure 3.3.

These results show several interesting features. At any point on the curve the value $d\gamma/d\ln C$ gives, from the Gibbs adsorption equation, (3.18), the corresponding value of the surfactant adsorption density or, alternatively, the surfactant head group or packing area at the water–air interface. As we will see in Chapter 8 another method for determining the surfactant head group area is afforded by the Langmuir trough technique. At surfactant concentrations just below the cmc value, the slope

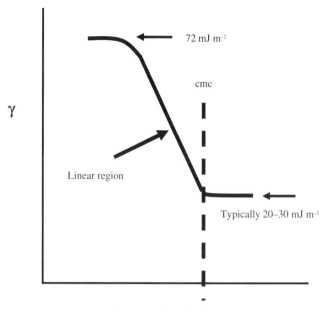

Figure 3.3 Typical graph of surface energy vs concentration for a micelle forming surfactant.

$d\gamma/d\ln C$ is linear, which from the Gibbs equation means that there is no further increase in the adsorption density with increase in bulk concentration. The surface is fully packed with surfactant molecules, although γ still continues to fall. This, apparently odd, situation arises because the chemical potential of the surfactant continues to increase with its concentration (see Equation 3.9) in this region and although its adsorption density does not change, this must reduce the energy required to create new surface, hence the surface energy continues to fall.

However, at the cmc a sharp transition occurs which apparently corresponds to zero adsorption (i.e. $d\gamma/d\ln C = 0$)! How can this be so? If we examine properties of the bulk solution in this region, we find that at this same concentration there is a sharp transition in a wide range of properties, such as conductivity, osmotic pressure and turbidity (see the following chapter). What is in fact happening is that the surfactant molecules are forming aggregates, usually micelles, and that all the additional molecules added to the solution go into these aggregates and so the concentration of monomers remains roughly constant. That is, both $d\gamma$ and $d\ln C$ are effectively zero, and the plot should strictly stop at the cmc value, since although we are adding surfactant molecules, we are not increasing their activity or concentration. The precise nature of these aggregates is discussed in the following chapter.

Industrial Report

Soil microstructure, permeability and interparticle forces

A soil is a condensed colloid system because the negatively charged, plate-shaped crystals are assembled in parallel or near-parallel alignment, to form stable operational entities, described as 'clay domains'. The crystals within a clay domain can be represented by a three-plate crystal model in which one crystal separates the other two crystals to produce a slit-shaped pore, where the crystals overlap. This situation is illustrated in Figure 3.4.

The surface separation in the slit-shaped pore is determined by the crystal thickness. For an illite (a fine-grained mica with a surface area of $1.6 \times 10^5 \, m^2$ per kg), the slit-shaped pores have a median size of about 5 nm and in the overlap pores the surface separation is about 1 nm. The stability of clay domains within a soil is a crucial feature for agricultural production because the permeability of a soil to aqueous electrolyte solutions depends on this stability. Swelling of these domains reduces permeability.

The interaction of clay crystals within a domain depends upon the DLVO repulsive pressure in the slit-shaped pores and the balance between repulsive pressure (P_R) from counterion hydration and the attractive pressure (P_A) generated by van der Waals forces and the recently discovered ion–ion correlation attraction between the counterions in the confined space of the overlap pores (see Kjellander et al., 1988a, b). When Ca^{2+} is the counterion, the attractive pressure dominates and the overlap pores are stabilized in a primary potential minimum. However, when the crystal

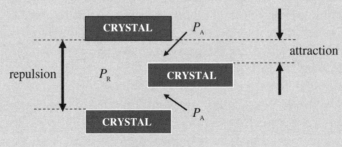

Figure 3.4

charge is balanced, in part, by Na^+ ions (called 'soil sodicity') the DLVO repulsive pressure in the slit shaped pores increases and the ion–ion correlation attraction is reduced. In dilute solutions the repulsive pressure, in the slit-shaped pores, is sufficient to release the platelets from the shallow potential minimum and the domains start to swell. However, in accord with DLVO theory, the swelling pressure, in the slit-shaped pores can be reduced by increasing the electrolyte concentration.

Studies on soils have shown that there is a nexus between saturated permeability (zero suction), sodicity and electrolyte concentration. The concentration, obtained by diluting the electrolyte, at which there is a first discernible decrease in permeability, called the 'threshold concentration', corresponds to the start of the swelling of the clay domains.

In irrigation agriculture this threshold concentration is used as a reference to adjust the concentration of irrigation water so that it exceeds the threshold concentration for the sodicity of the soil and so prevents decreases in permeability. The threshold concentration increases with the degree of sodicity (see Quirk, 2001).

At about one-quarter of the threshold concentration, for a given sodicity, dispersed particles appear in the percolate, indicating the start of the dismantling of clay domains. It is noteworthy that this concentration is almost ten times lower, or even more if natural dispersants are present (e.g. organic compounds), than that obtained for the flocculation of a suspension of the soil. This reflects the fact that it is harder to release the crystals from within the clay domains, than to simply flocculate the free crystals.

Professor J.P. Quirk
Formerly Director
Waite Agricultural Research Institute
Adelaide, Australia

Sample problems

1. The surface tension data given in Figure 3.5 was obtained for aqueous solutions of a trivalent cationic surfactant ($CoRCl_3$) in both water and in 150 mM NaCl solution. Use the data and the Gibbs adsorption isotherm to obtain estimates of the minimum surface area per molecule adsorbed at the air/water interface.

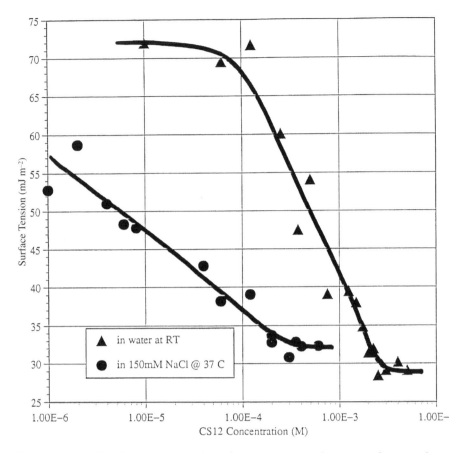

Figure 3.5 Surface tension data for aqueous solutions of a trivalent surfactant $CoRCl_3$.

2. Use the Gibbs adsorption isotherm to describe the type of surfactant adsorption occurring at the air/water interface at points A, B, C and D in Figure 3.6.

Experiment 3.1　Adsorption of acetic acid on to activated charcoal

Introduction

Activated charcoal or carbon is widely used for vapour adsorption and in the removal of organic solutes from water. These materials are used in industrial processes to purify drinking water and swimming pool water, to de-colorize sugar solutions as well as other foods, and to

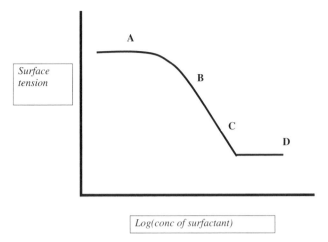

Figure 3.6 Schematic diagram of the decrease in surface tension for a typical surfactant solution.

extract organic solvents. They are also used as a first, oral, treatment in hospitals for cases of poisoning. Activated charcoal can be made by heat degradation and partial oxidation of almost any carbonaceous material of animal, vegetable or mineral origin. For convenience and economic reasons it is usually produced from bones, wood, lignite or coconut shells. The complex three-dimensional structure of these materials is determined by their carbon-based polymers (such as cellulose and lignin) and it is this backbone which gives the final carbon structure after thermal degradation. These materials, therefore, produce a very porous high-surface-area carbon solid. In addition to a high area the carbon has to be 'activated' so that it will interact with and physisorb (i.e. adsorb physically, without forming a chemical bond) a wide range of compounds. This activation process involves controlled oxidation of the surface to produce polar sites. In this experiment we will examine quantitatively the adsorption properties of a typical granular charcoal. Adsorption at liquid surfaces can be monitored using the Gibbs adsorption isotherm since the surface energy of a solution can be readily measured. However, for solid subtrates this is not the case and the adsorption density has to be measured in some other manner. In the present case the concentration of adsorbate in solution will be monitored. In place of the Gibbs equation we can use a simple adsorption model based on the mass action approach.

If we assume that the granular charcoal has a certain number of possible adsorption sites per gram (N_m) and that a fraction θ are filled by the adsorbing solute, then:

$$\text{the rate of adsorption } \alpha \text{ [solute conc.]}[1 - \theta]N_m$$

and

$$\text{the rate of desorption } \alpha \ \theta N_m$$

At equilibrium these rates must be equal, hence

$$k_a C(1 - \theta)N_m = k_d \theta N_m \qquad (3.19)$$

where k_a, k_d are the respective proportionality constants and C is the bulk solution concentration of solute. If we let $K = k_a/k_d$, then

$$C/\theta = C + 1/K \qquad (3.20)$$

and since $\theta = N/N_m$, where N is the number of solute molecules adsorbed per gram of solid, we obtain the result that

$$\frac{C}{N} = \frac{C}{N_m} + \frac{1}{KN_m} \qquad (3.21)$$

Thus, measurement of N for a range of concentrations (C) should give a linear plot of C/N against C, where the slope gives the value of N_m and the intercept the value of the equilibrium constant K.

This model of adsorption was suggested by Irving Langmuir and is referred to as the 'Langmuir adsorption isotherm'. The aim of this experiment is to test the validity of this isotherm equation and to measure the surface area per gram of charcoal, which can easily be obtained from the measured N_m value, if the area per solute molecule is known.

Experimental details

In this experiment it is important to measure the acetic acid concentrations accurately. To this end the NaOH solution must be titrated with standard 0.1 M HCl solution and then titrated with the acetic acid solution. (Question: Why is NaOH solution not used as a standard?)

Weigh out 1 g of granulated activated charcoal into each of five clean, stoppered 250 cm³ conical flasks. Add 100 cm³ of 0.2 M acetic acid (stock solution) to the first flask and shake. Then add $100\,cm^3$ to each of the other flasks at concentrations of: 0.15, 0.10, 0.07 and 0.03 M. Shake each loosely stoppered flask periodically over 30 minutes. Then allow them to stand for 1 hour, noting the room temperature. With-draw just over $50\,cm^3$ of the solution and filter through a fine sinter (to completely remove charcoal particles) and titrate two $25\,cm^3$ portions with 0.1 and 0.01 M NaOH (depending on the initial acid concentra-tion). As indicator use phenolphthalein. It is important that the equi-librium acetic acid concentration be accurately determined.

Calculate the number of acetic acid molecules adsorbed per gram of charcoal (N) and the corresponding equilibrium acid concentration (C). Plot N against C and C/N against C. Determine the surface area per gram of charcoal assuming that one adsorbed acetic acid molecule occupies an area of $21\,Å^2$. Estimate the value of the equilibrium con-stant K with the correct units.

Questions

1. The area per gram of activated charcoal as determined by nitrogen adsorption is typically in the range 300–$1000\,m^2/g$. Why is this value different from that determined in this experiment?

2. Why should the carbon surface be activated by oxidation and what would you expect the typical surface groups to be?

3. Why is a slurry of charcoal in water given orally to suspected poison victims?

4

Surfactants and Self-Assembly

Introduction to the variety of types of surfactants, effect of surfactants on aqueous solution properties. Law of mass action applied to the self-assembly of surfactant molecules in water. Spontaneous self-assembly of surfactants in aqueous media. Formation of micelles, vesicles and lamellar structures. Critical packing parameter. Detergency. Laboratory project on determining the charge of a micelle.

Introduction to surfactants

The name 'surfactant' refers to molecules that are 'surface-active', usually in aqueous solutions. Surface-active molecules adsorb strongly at the water–air interface and, because of this, they substantially reduce its surface energy (Gibbs theorem). This is the opposite behaviour from that observed for most inorganic electrolytes, which are desorbed at the air interface and hence raise the surface energy of water (slightly). Surfactant molecules are amphiphilic, that is, they have both hydrophilic and hydrophobic moieties, and it is for this reason that they adsorb so effectively at interfaces (note that 'amphi' means 'of both kinds' in Greek).

Applied Colloid and Surface Chemistry Richard M. Pashley and Marilyn E. Karaman
© 2004 John Wiley & Sons, Ltd.   ISBN 0 470 86882 1 (HB)   0 470 86883 X (PB)

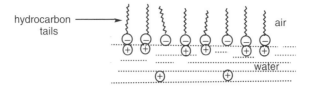

Figure 4.1   Schematic diagram of surfactant molecules adsorbed at the water/air interface.

'Natural' surfactants, such as soaps, are made by saponification of fats or triglycerides, such as tri-palmitin in palm oil. The main component of common soap is sodium stearate, $C_{17}H_{35}COO^- Na^\oplus$, which is made from the saponification of animal fats. When dissolved in water, the carboxylic headgroup ionizes and is strongly hydrophilic, whereas the hydrocarbon chain is hydrophobic. The hydrocarbon chain, alone, is almost completely insoluble in water. When dissolved into aqueous solution, the molecules can adsorb and orientate at the air/solution interface, as illustrated in Figure 4.1, to reduce the surface tension of water:

The interfacial energy is typically reduced down to about 30–35 mJ m^{-2}, and the surface now appears more like that of a hydrocarbon. The surfactant molecules are in a lower energy state when immersed in bulk water, than when adsorbed at the surface. However, the displacement of even less favourable water molecules from the air surface dominates the overall process and the surfactant molecules are preferentially adsorbed. It is the hydrocarbon tail which makes the molecule less favourable in water. The methylene groups can neither hydrogen-bond nor form dipole bonds with water. Water molecules surrounding the hydrocarbon groups therefore orientate or order, so as to increase the number of bonds to other neighbouring water molecules. This increase in local order (decreasing entropy) increases the free energy of these water molecules relative to those in bulk solution.

A list of typical surfactant molecules, with different types of charge, is given in Table 4.1.

Common properties of surfactant solutions

In addition to the surface adsorption properties of surfactants, they also have the remarkable ability to self-assemble in aqueous solution. The structures spontaneously formed by surfactants in solution are created

Table 4.1

ANIONIC:
Sodium dodecyl sulphate (SDS) $CH_3(CH_2)_{11}SO_4^-Na^+$
Sodium dodecyl benzene sulphonate $CH_3(CH_2)_{11}C_6H_4SO_3^-Na^+$

CATIONIC:
Cetyltrimethylammonium bromide (CTAB) $CH_3(CH_2)_{15}N(CH_3)_3^+Br^-$
Dodecylamine hydrochloride $CH_3(CH_2)_{11}NH_3^+Cl^-$

NON-IONIC:
Polyethylene oxides   e.g. $CH_3(CH_2)_7(O.CH_2CH_2)_8OH$
                                                (called C_8EO_8)

ZWITTERIONIC:

Dodecyl betaine $C_{12}H_{25}N^+ \big\langle \begin{matrix} (CH_3)_2 \\ CH_2COO^- \end{matrix}$

Lecithins, e.g. phosphatidyl choline

$$\begin{matrix} & O \\ & \| \\ CH_2OCR \\ | & O \\ & \| \\ CHOCR \\ | & O \\ & \| & + \\ CH_2OP\text{-}O\text{-}CH_2\text{-}N(CH_3)_3 \\ | \\ O^- \end{matrix}$$

to reduce the exposure of the hydrocarbon chains to water. Many of their solution properties reflect this ability, as is illustrated in Figure 4.2, which shows typical solution behaviour for single-chained ionic surfactants, such as CTAB and SDS.

A sharp transition occurs in most of the solution properties, which corresponds to the formation of self-assembled structures called 'micelles'. The concentration at which they are formed is a characteristic of the particular surfactant and is called the 'critical micelle concentration', or cmc. A section through a micelle would look something like Figure 4.3. These small structures typically contain about 100 molecules, with the central core of the micelle essentially a water-free, liquid hydrocarbon environment. The formation of aggregates is a very important property of surfactants and is of fundamental importance in their detergent cleaning action. The hydrocarbon regions in the aggregates solubilize fatty, organic materials during cleaning. This organic 'dirt' is otherwise insoluble in water. As an example, if liquid paraffin is stirred into a soap solution, the solution remains clear until the

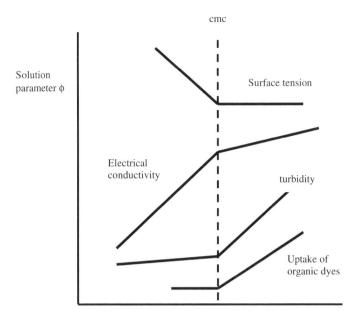

Figure 4.2 Diagram illustrating the sharp change in a range of solution properties at the cmc.

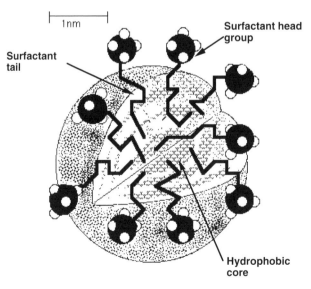

Figure 4.3 Schematic diagram of a surfactant micelle.

Figure 4.4 Sudan yellow is a water-insoluble dye, seen at the bottom of the cylinder on the left but fully dissolved in the micelle solution on the right.

capacity of the micelles to absorb paraffin is exceeded. Water-insoluble, coloured organic dyes, such as Sudan yellow, are clearly taken in micellar solutions, as illustrated in Figure 4.4.

Thermodynamics of surfactant self-assembly

That surfactant molecules form aggregates designed to remove unfavourable hydrocarbon–water contact is not surprising but the question that should be asked is why the aggregates form sharply at a concentration characteristic of the surfactant (the cmc). From the basic equation of ideal solution thermodynamics

$$\mu_i = \mu_i^0 + kT \ln x \qquad (4.1)$$

it is clear that as we increase the concentration of the surfactant, x the mol fraction increases, and so does the chemical potential of the molecules. (Note that this equation was written earlier, in Chapter 3, in terms of molar concentrations, with an appropriate adjustment in standard state.) If we make the pseudo-phase approximation that the micelles can be considered as a species like the surfactant monomers, we also obtain a similar relationship that

$$N\mu_N = N\mu_N^0 + kT\ln(x_N/N) \tag{4.2}$$

where $N\mu_N$ is the chemical potential of an aggregate of N surfactant molecules and x_N is the mole-fraction of surfactant molecules in the N-aggregates. Now, at equilibrium $\mu_1 = \mu_N$, that is, the chemical potentials of free monomers and monomers in the aggregates must be the same. Thus, we obtain the result that

$$\mu_i^0 + kT\ln x_1 = \mu_N^0 + [kT\ln(x_N/N)]/N \tag{4.3}$$

This can be rearranged to show that

$$\frac{x_1^N}{x_N} = \text{a constant} \quad \text{(for a fixed value of } N) \tag{4.4}$$

This is simply the result we would expect from the law of mass action. Thus, for the reaction: N monomers $\Leftrightarrow$ 1 micelle, we can immediately write that there is an equilibrium (association) constant given by

$$\frac{x_{\text{mic}}}{x_1^N} = K_e \tag{4.5}$$

or

$$\frac{x_N}{Nx_1^N} = K_e \tag{4.6}$$

Now, if we define the cmc as that mole fraction of surfactant at which $x_N \cong x_1 = x_{\text{cmc}}$, then, at the cmc,

$$\frac{x_{\text{cmc}}^{(1-N)}}{N} = K_e \tag{4.7}$$

In addition, if this definition of the cmc is also incorporated into (4.3), then we obtain the result that for large values of N (e.g. 100)

$$\ln(x_{\text{cmc}}) = \frac{\mu_N^0 - \mu_1^0}{kT} \tag{4.8}$$

Equations (4.7) and (4.8) are both very useful when analysing surfactant aggregation behaviour and experimental cmc values. The difference in standard chemical potentials $\mu_N^0 - \mu_1^0$ must contain within it the molecular forces and energetics of formation of micelles, which could be estimated from theory. These will be a function of the surfactant molecule and will determine the value of its cmc.

The above analysis actually refers to the simplest case of aggregation of non-ionic surfactants. For monovalent, ionic surfactants the aggregation reaction becomes:

$$N \text{ ionic monomers} + Q \text{ counter-ions} \Leftrightarrow \text{micelle}$$

because the ionic monomers will usually be fully ionized but the high electric field strength at the surface of the micelles will often cause adsorption of some proportion of the free counterions. Thus, the micelles will have a total ionic charge of $(\pm)(N - Q)$. The equilibrium constant is, in this case, given by

$$K_e = \frac{x_{\text{mic}}}{x_1^N x_{\text{ion}}^Q} \tag{4.9}$$

Calculated concentrations, using (4.9), for the various components, surfactant monomers, counter-ions and micelles, for the case of CTAB micellization (with a cmc of 0.9 mM), is shown in Figure 4.5. Clearly, the micelle concentration increases rapidly at the cmc, which explains the sharp transition in surfactant solution properties referred to earlier. It is also interesting to note that the law of mass action (in the form of equation 4.9) predicts an increase in counterion (Br^- ions) concentration and a decrease in free monomer concentration above the cmc. It has been proposed that for ionic surfactants, a useful definition of the cmc would be

$$\frac{dn_1}{dc} = 0 \text{ at the cmc}$$

rather than the usual experimental definition of

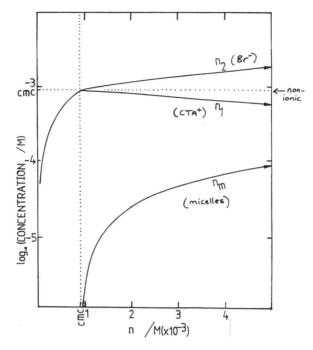

Figure 4.5 Calculated concentration of micelles, CTA$^+$ and Br$^-$ ions for CTAB solution near the cmc.

$$\frac{d^3\phi}{dc^3} = 0$$

which is difficult to use theoretically.

The equivalent curves for non-ionic surfactants, obtained using (4.6) and (4.7), give a constant monomer concentration above the cmc, with a similar increase in micelle concentration.

Self-assembled surfactant structures

Ionic surfactants actually only form micelles when their hydrocarbon chains are sufficiently fluid, that is at temperatures above their chain melting temperature. Below a specific temperature for a given surfactant, the Krafft temperature, the surfactant becomes insoluble rather than self-assembles. For CTAB this temperature is around 20 °C and only above this temperature are micelles formed. In general, the longer the hydrocarbon chain length, the higher the Krafft temperature. For this reason, shorter-chain-length surfactants or branched chain soaps are used for cold-water detergent formulations. Non-ionic surfactants suffer the inverse problem and become insoluble with increasing temperature. This is because their polar head groups become less hydrophilic with increasing temperature and a 'cloud point' is reached where they precipitate from solution.

Micelles are not the only self-assembled structures that can be formed. The physical constraints on the surfactant molecule dictate the type of aggregate which it can form to exclude water. Possible structures can be modelled simply by using the surfactant's calculated hydrocarbon chain volume and its head-group area. These simple ideas can be used to illustrate clearly why double-chain surfactants, such as the lecithins, do not form micelles but instead form vesicles, liposomes and multi-lamellar bilayers. These remarkable aggregates closely mirror the kinds of structures observed in living cell membranes. The chain volume varies with the surfactant type but the addition of suitable chain-penetrating oils can increase this volume. An ionic surfactant head-group area can also be varied by the addition of screening electrolyte, which has the effect of reducing the head-group area and hence can dramatically change its aggregation properties. In addition, co-surfactants (usually long-chain alcohols) can be incorporated to change the packing. These can also lower the minimum surface energy at the air/solution interface.

Molecular organization or self-assembly depends upon a number of competing intramolecular forces, the flexibility of the chains and the intermolecular forces. The relative magnitudes of the attractive hydrophobic forces between the hydrophobic tails, the repulsive electrostatic forces between the charged head groups and the head-group hydration effects all influence aggregate architecture and stability. Originally it was thought that in order to predict the physical characteristics of an aggregate (e.g. size, shape) it would be necessary to have detailed knowledge of the complex intermolecular forces acting between the polar head groups and their hydrocarbon tails. However, it soon became clear that simple packing constraints offered a valuable tool for the prediction of aggregate structure. This view of aggregation saw the emergence of a relatively simple characterization of self-assembly based on the degree of curvature existing at the aggregate surface. This curvature can be expressed as a dimensionless parameter known as the "critical packing parameter" $v/a_0 l_c$, where v is the volume of the hydrocarbon chain (assumed to be fluid and incompressible) given below, l_c is the critical chain length, assumed to be approximately equal to l_{max}, the fully extended chain length, n is the number of carbon atoms in the hydrocarbon chain, m is the number of hydrocarbon chains, and a_o is the head-group area (which can be estimated using the Langmuir trough technique, see later):

$$v = (27.4 + 26.9n)m \quad \left(\text{units, Å}^3\right)$$

$$l_{\max} = 1.5 + 1.265n \quad (\text{units, Å})$$

The critical packing parameter can be used as a guide to the aggregate architecture for a given surfactant (as shown in Figure 4.6). Typical values and their corresponding aggregate structures are:

$v/a_o l_c < 1/3$ Spherical micelles

$1/3 < v/a_o l_c < 1/2$ Poly-dispersed cylindrical micelles

$1/2 < v/a_o l_c < 1$ Vesicles, oblate micelles or bilayers

$v/a_o l_c > 1$ Inverted strucutres

Earlier theories dealt with only these simple shapes; however, more complicated structures (e.g. "cubic" and other bicontinuous phases) predicted by geometric packing arguments, have since been confirmed. The critical packing parameter is a useful parameter in aggregate design, as it can be changed for a given ionic surfactant by the addition of electrolyte, addition of co-surfactant, change in temperature, change in counterion, or insertion of unsaturated or branched chains. Some of the basic structures are illustrated in Figure 4.6. See, also, the important text *The Hydrophobic Effect* (Tandon 1980). Controlling aggregate architecture has enormous potential in many new areas of biochemical research, catalysis, drug delivery and oil recovery, to name but a few.

Surfactants and detergency

In the detergency process, fatty materials (i.e. dirt, often from human skin) are removed from surfaces, such as cloth fibres, and dispersed in water. It is the surfactants in a detergent which produce this effect. Adsorption of the surfactant both on the fibre (or surface) and on the grease itself increases the contact angle of the latter as illustrated in Figure 4.7. The grease or oil droplet is then easily detached by mechanical action and the surfactant adsorbed around the surface of the droplet stabilises it in solution.

Lipid	Critical packing parameter $\frac{v}{a_o l_c}$	Critical packing shape	Structures formed
Single-chained lipids with large head-group areas. (e.g. NaDS in low salt and some lysophospholipids)	$< \dfrac{1}{3}$	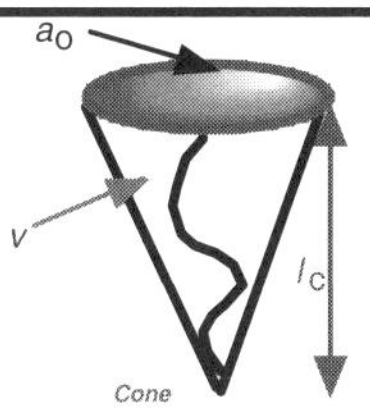 Cone	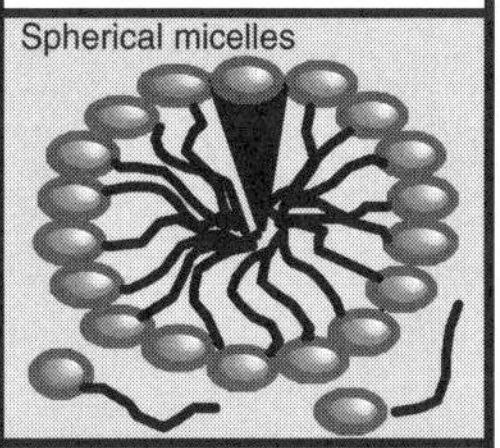 Spherical micelles
Single-chained lipids with small head-group areas. (e.g. NaDS in high salt, lysolecithin and non-ionic surfactants)	$\dfrac{1}{3} - \dfrac{1}{2}$	 Truncated cone or wedge	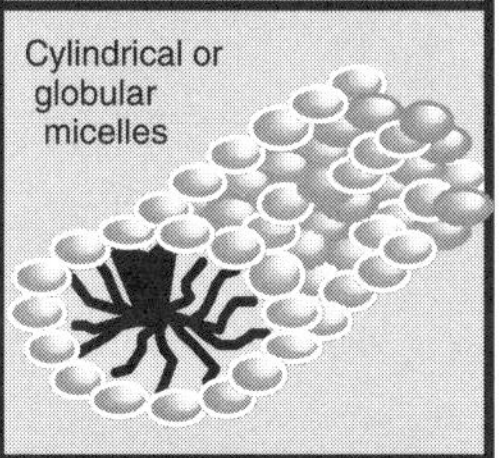 Cylindrical or globular micelles
Double-chained lipids with large head-group and fluid chains. (e.g. lecithin, dialkyl dimethyl ammonium salts, sphingomyelin, DGDG, phosphatidylserine, phosphatidyl inositol)	$\dfrac{1}{2} - 1$	 Truncated cone	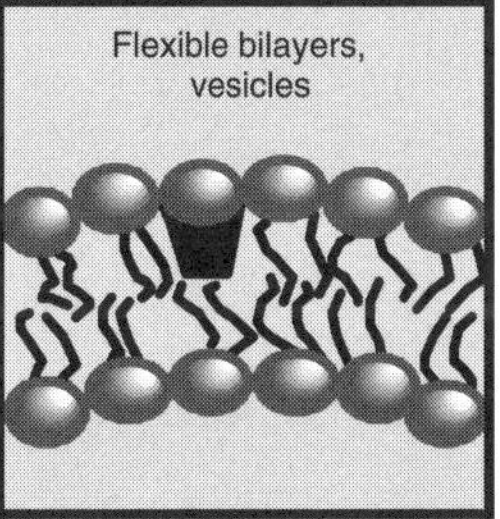 Flexible bilayers, vesicles
Double-chained lipids with small head-group areas: anionic lipids high salt, saturated frozen chains. (e.g. phosphatidyl ethanolamine, phosphatidyl serine + Ca^{2+})	~1	 Cylinder	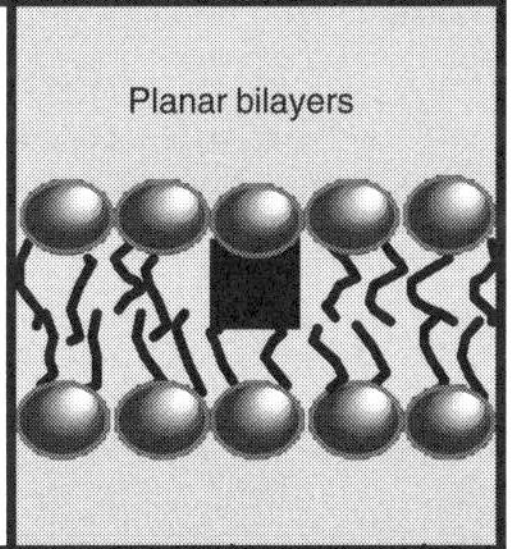 Planar bilayers
Double-chained lipids having small head groups (e.g. non-ionic lipids, poly(cis) unsaturated chains, phosphatidic acid + Ca^{2+})	> 1	 Inverted truncated cone	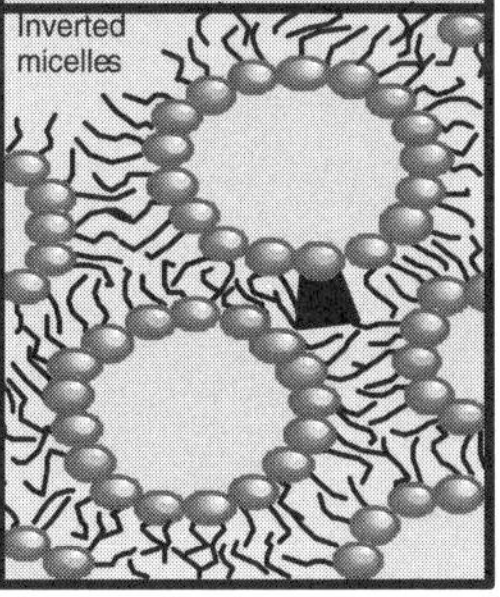 Inverted micelles

Figure 4.6 Use of the critical packing parameter to predict aggregate structures.

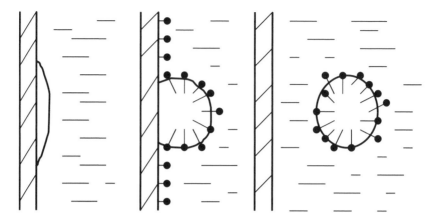

Figure 4.7 Illustration of the removal of hydrophobic oil from a fibre using detergent.

The composition of a typical powdered laundry detergent is given below:

Na alkylbenzenesulphonate	15%
Anhydrous soap	3%
Sodium tripolyphosphate or powdered zeolite	30%
Sodium silicate	14%
Sodium carbonate	10%
Sodium sulphate	18%
Sodium carboxymethyl cellulose	1%
Optical brighteners, perfume, moisture	~0

The first two components are the 'active' surfactants, whereas the other components are added for a variety of reasons. The polyphosphate chelate Ca^{2+} ions which are present (with Mg^{2+} ions also) in so-called 'hard waters' and prevents them from coagulating the anionic surfactants. Zeolite powders are often used to replace phosphate because of their nutrient properties in river systems. Sodium silicate is added as a corrosion inhibitor for washing machines and also increases the pH. The pH is maintained at about 10 by the sodium carbonate. At lower pH values the acid form of the surfactants are produced and in most cases these are either insoluble or much less soluble than the sodium salt. Sodium sulphate is added to prevent caking and ensures free-flowing powder. The cellulose acts as a protective hydrophilic sheath around dispersed dirt particles and prevents re-deposition on the fabric. Foam stabilizers (non-ionic surfactants) are sometimes added to give a

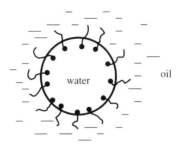

Figure 4.8 Diagram of how surfactant molecules stabilise water droplets in oil.

visible signal that sufficient detergent has been added. However, the creation of a foam is not necessary for detergent action in a conventional washing machine. A lather is, however, of value for personal washing, as it provides a mobile concentrated soap solution on the skin. In the 1970s the widespread use of non-biodegradable synthetic detergents led to extensive foaming in rivers and lakes and the consequent death of aquatic life through lack of aeration of the water. Great care is now taken to use readily biodegradable soaps.

An inverted type of detergency process is used in dry-cleaning, where a non-aqueous liquid, usually tetrachloroethylene, is used to dissolve grease from materials that should not be exposed to water. However, because complete cleaning must also remove polar materials, for example sugars which are insoluble in C_2Cl_4, surfactants and a small amount of water are added, forming inverted micelles as in Figure 4.8. Polar materials are then solubilized inside the aqueous core of the micelle and the material is not exposed to water, even though the process is not entirely dry. Mixtures of oils with different surfactants can be used to produce sticky, inflammable gels such as napalm made from palmitic soaps and aluminium salts. When water is replaced by a concentrated solution of a soluble oxidant, such as ammonium nitrate, the intimate contact with a reactant, such as hydrocarbon oil, can be used as the basis of a 'plastic' explosive.

Finally, surfactants have also been used to reduce water evaporation from open reservoirs in arid areas, especially in Australia. The packed insoluble monolayer adsorbed at the air/water interface substantially reduces the transfer of water vapour to the atmosphere. Cetylalcohol is used at the rate of 1 ounce per acre per day for this reason. It has been calculated that this procedure can save up to one million gallons per acre per year.

Industrial Report

Colloid science in detergency

The removal of particulate soil from a fabric surface is a key step in the overall detergency process with the balance between van der Waals and electrical double-layer forces playing a major role. However, once the particle has been removed from the fabric surface keeping it from redepositing at some other stage in the process is vital if the article is to remain clean at the end of the wash process. Consequently, one needs to ensure that a barrier to deposition is established that will persist ideally throughout the process. So to ensure that redeposition does not occur polymeric anti-redeposition agents (ARDs) are often added to formulation. These ARD polymers work by adsorbing on both the substrate and the particulate soil, thereby creating a protective layer that can both sterically and electrostatically hinder redeposition of the previously removed soil. Typical ARD polymers are anionics such as sodium carboxymethyl cellulose (SCMC) and non-ionic cellulose ethers such as methylhydroxypropylcellulose (MHPC). Anionic polymers are particularly suitable for use with hydrophilic fabrics such as cotton and cotton mix whereas non-ionics can be used on both polyester and cotton fabrics. Both types of polymer function by altering the surface properties of the fibres to render them more hydrophilic and by adsorbing at the particle and fabric surfaces to present a barrier via the absorbed layer to redeposition. SCMC is specific to hydrophilic fibres such as cotton and has best activity when the degree of substitution is below 0.7. In contrast, the non-ionic MHPC has a broader spectrum of activity, being effective on both hydrophilic and especially hydrophobic fibres such as polyester. Of course, in real detergent systems both surfactants and ARD polymers compete for the fabric and particle surfaces and this competition can under certain circumstances reduce the polymer's effectiveness. This competition is seen to be most problematic with non-ionic surfactants as they compete for surface sites with the polymer. To ensure effectiveness under such conditions the formulator needs to balance the surfactant and polymer levels carefully. Typical polymer levels are around 0.5% to 1.0% w/w in the formulation. Other polymer types that have been explored as ARDs are polyethylene/

polyoxyethylene terephthalate co-polymers and hydrophobically modified polyethylene glycol.

Dr Ian C. Callaghan
Head of Formulation Technology
Lever Faberge Europe – Global Technology Centre
Unilever Research
Vlaardingen
The Netherlands

Sample problems

1. Describe how the critical packing parameter for surfactant self-assembly can be used to describe the structure of typical biological lipid membranes.

2. Explain the link between the critical packing parameter and the interaction forces between surfactant molecules in water.

3. Use the cmc values of a homologous series of single-chained sodium sulphate surfactants, given below, to estimate the standard free energy of transfer of a methylene (-CH$_2$-) group from an aqueous to a hydrocarbon environment.

Number of carbons in tail	12	14	16	18
cmc/mM	8.6	2.2	0.58	0.23

Experiment 4.1
Determination of micelle ionization

Introduction

In this experiment the degree of ionization of CTAB micelles is determined by measuring the change in slope of solution electrical conductivity (κ) versus total concentration (C) as the solution goes through the critical micelle concentration (cmc). That this information is sufficient to estimate the degree of ionization of the micelles formed at the cmc can be shown by the following simple analysis.

First of all, let us make the assumption that the overall solution conductivity (κ) is entirely due to the free Br^- ions present in solution. That is, we assume that the conductivity due to the much larger CTA^+ ions and the even larger micelles is negligible because of the size of these species (and hence large viscous drag). Thus we assume that

$$\kappa \;\; \alpha \;\; C_{Br^-} \tag{4.10}$$

Hence, at CTAB concentrations (C) below the cmc ($C <$ cmc)

$$\kappa = AC \tag{4.11}$$

since the CTAB monomer is always fully ionized and A is some constant. By comparison, at concentrations above the cmc ($C >$ cmc) it must also be true that

$$C_{Br^-} = \text{cmc} + \alpha(C - \text{cmc}) \tag{4.12}$$

where α is the degree of ionization of the micelle, which is assumed to be independent of concentration. Thus, above the cmc

$$\kappa = A[\text{cmc} + \alpha(C - \text{cmc})] \tag{4.13}$$

Now, from (4.11) and (4.13) we can calculate the gradients above and below the cmc:

$$\left.\frac{d\kappa}{dc}\right|_{c>\text{cmc}} = \alpha A \tag{4.14}$$

and

$$\left.\frac{d\kappa}{dc}\right|_{c<\text{cmc}} = A \tag{4.15}$$

and so, obviously, the ratio of these slopes gives the degree of ionization of the micelles.

Experimental details

All electrical conductivity measurements should be carried out on solutions equilibrated in a thermostat bath at 25 °C. Conductivity values (i.e. κ in units of ohm^{-1}cm^{-1}) should be determined at 1000 Hz and over a concentration range of 0.0003 M to 0.003 M, taking at least four measurements on either side of the cmc (~0.001 M). Plot a linear graph of the results and determine the cmc of CTAB and the approximate degree of ionization of the micelles. Note that in using surfactant solutions, one should always try to prevent foaming by not shaking solutions too vigorously.

Questions

(1) Discuss the main assumptions used in the method described here to estimate the degree of ionization of micelles.

(2) What is the total charge on a CTAB micelle if the head-group area is $45\,\text{Å}^2$ and the micelle is $40\,\text{Å}$ in diameter?

5
Emulsions and Microemulsions

The conditions required to form an emulsion of oil and water and a microemulsion. The complex range of structures formed by a microemulsion fluid. Emulsion polymerization and the production of latex paints. Photographic emulsions. Emulsions in food science. Laboratory project on determining the phase behaviour of a microemulsion fluid.

The conditions required to form emulsions and microemulsions

Adsorption of a surfactant monolayer at the air/solution interface still produces a surface with a significant interfacial energy, typically of between 30 and $40 \, \mathrm{mJ \, m^{-2}}$, because the surface is now similar to that of a typical hydrocarbon liquid. However, in comparison, adsorption at a water/oil interface offers the opportunity of creating surfaces with very low interfacial energies, which ought to produce some interesting effects and possibilities. The formation of this type of low interfacial energy surface is the basis of the stability of most oil and water emulsions and all microemulsions. This situation is illustrated in Figure 5.1.

Applied Colloid and Surface Chemistry Richard M. Pashley and Marilyn E. Karaman
© 2004 John Wiley & Sons, Ltd.   ISBN 0 470 86882 1 (HB)   0 470 86883 X (PB)

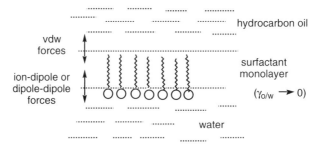

Figure 5.1 Illustration of the effect of an adsorbed surfactant layer on the interfacial energy between oil and water.

It is a well-known observation that oil and water do not mix. Fundamentally, this is because hydrocarbon molecules are non-polar and cannot interact strongly with water molecules, which have to be forced apart to incorporate the hydrocarbon solute molecules. Hence, hydrocarbon oils will not dissolve in water. However, when oil and water are vigorously shaken together a droplet emulsion can be formed. This mixture will destabilize fairly quickly and phase-separate into oil and water, typically within less than an hour, because of the high interfacial energy of the oil–water droplets. The stability of these emulsions can be substantially enhanced by the addition of surfactants which will reduce the interfacial energy of the droplets. Emulsion droplets of oil-in-water or water-in-oil typically fall in the range of 0.1–10 μm, whereas microemulsion droplets are in the 0.01–0.1 μm size range. However, in the latter case the actual structure of the phase is still very controversial (see later). For emulsions the $\gamma_{o/w}$ value is typically in the range 0.1–1 mJ m^{-2} and for microemulsions $\gamma_{o/w}$ is as low as 0.001 mJ m^{-2}. The addition of emulsifying agents (usually surfactant mixtures) produces either an opaque stable emulsion or a clear microemulsion. This method of solubilizing oil in water is used in foodstuffs (dairy produce), agricultural sprays and pharmaceutical preparations.

Lowering of the interfacial energy enables the formation of high-surface-area emulsions but additional factors are also involved in preventing droplet collisions from causing a phase separation (compare this situation with the coagulation of sols in Chapter 7). These factors include electrical repulsion for charged emulsifiers, polymers such as proteins adsorbed to give mechanical prevention of film drainage (and hence droplet coalescence), finely divided solid particles (such as clays and carbons) adsorbed around the interface, and an increased viscosity. The type of emulsifier used determines whether an oil-in-water or a water-in-oil emulsion is formed. An empirical numbering system has been developed to enable the correct type of surfactant to be chosen. The system is called the 'hydrophile–lipophile balance', HLB. The most hydrophilic surfactants have the highest HLB values. In general, the phase in which the emulsifying agent is more soluble tends to be the dispersion medium.

Emulsions are metastable systems for which phase separation of the pure oil and water phases represents the most stable thermodynamic state. However, microemulsions, in which the interfacial energies approach zero, may be thermodynamically stable. Also, the microemulsion phase is clear, which indicates very small droplets and a very high interfacial area. In fact, it is not at all certain what is the precise structural nature of many microemulsions. It has been postulated that a range of intricate bicontinuous structures may exist, rather than simple droplets like swollen micelles. Whatever the nature of microemulsions, there is currently a large international effort towards understanding these phases. One of the most important reasons for this is because of the possibility of using them to increase the yield from vast oil reserves in capillary rocks, which at present cannot be tapped. As an example, something like 95 per cent of the anticipated oil wells have already been discovered in the USA, but the overall average recovery is less than 40 per cent; clearly a large amount of oil has yet to be removed.

The possible structures which can be formed by a mixture of hydrocarbon oil, surfactant and water is illustrated below in Figure 5.2. The variation and complexity of these structures has led to much research on potential industrial applications from tertiary oil recovery to enhanced drug delivery systems. Many of the structures can be predicted using models based on the optimal curvature of the interface, not unlike that used to predict surfactant aggregation.

Emulsion polymerization and the production of latex paints

The aim of the modern 'emulsion' painting process is to deposit a uniform, tough polymer layer on a substrate. At first sight it may be thought that simply dissolving the polymer (typically polyacrylic) in a suitable non-aqueous solvent would be sufficient. However, polymer solutions of the required concentrations (~50%) are very viscous and organic solvents are not acceptable for the home decoration market as well as being expensive.

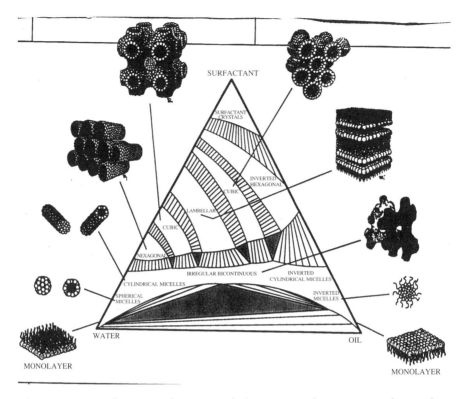

Figure 5.2 Schematic diagram of the types of structures formed at different compositions of oil, water and surfactant.

Albert Einstein derived a simple equation for the viscosity of a solution of spherical particles, and from this result it is obvious that if we could make the polymer in small colloidal-sized balls, then the solution would be much less viscous. Also, if we could use surfactants to stabilize (e.g. by charging) the polymer particles in water, then there would be no need for organic solvents. Both these conditions are neatly obtained in the emulsion polymerization process, which is schematically explained in Figure 5.3. A polymer 'latex' is produced by this process and can contain up to 50% polymer in the form of 0.1–0.5 μm size spherical particles in water. A typical starting composition is:

Monomer	50%	In commercial latex dispersions this is often a mixture of acrylic acid and butyl and ethyl acrylates.
Soap	2%	

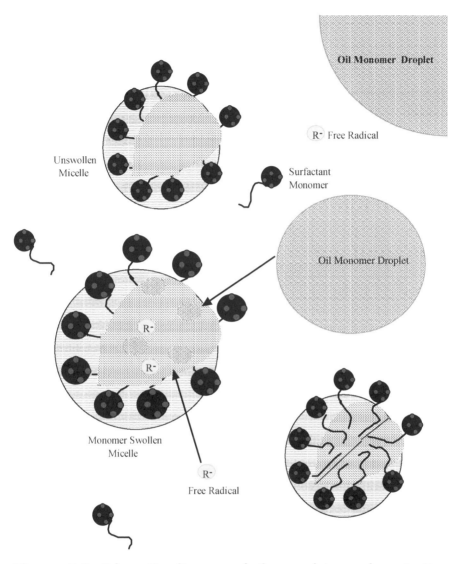

Figure 5.3 Schematic diagram of the emulsion polymerization process.

$C_{12}H_{25}SH$	0.2%	This is a terminator or chain transfer agent to reduce the molecular weight of the resulting polymer.
$K_2S_2O_8$	0.1%	This is a thermal radical initiator that generates sulphate radicals.
Water	47–48%	

The final 'emulsion' paint is produced by adding pigments (which are also colloidal particles), antimould agents, wetting agents and plasticizer to the latex. After spreading the paint onto a surface the water evaporates and draws the particles together until they fuse into a uniform polymer layer with embedded pigment (see Chapter 7). It should be noted that the term 'emulsion paint' is not strictly correct because these latex paints are solid, although soft, dispersions in water – they are colloidal solutions. This is even more the case for photographic 'emulsions'.

Photographic emulsions

The photographic 'emulsions' used in both black-and-white and colour photography are essentially colloidal solutions of AgBr crystals (0.02–4 μm) in a gel of gelatin and water. On heating, the gel becomes liquid-like and at this stage NH_4Br is added to $AgNO_3$ in the solution to give a dispersion of insoluble AgBr crystals which are fixed in the gel matrix on cooling. The 'emulsion' is coated in a 25 μm layer on a transparent acetyl plastic base to form the film. The fundamental process on exposure to light is the production of a latent image. In this process a few surface Ag^+ ions on some of the crystals are converted to Ag^0 atoms. These are not enough to see but act as nuclei for further, much more substantial, conversion to Ag^0 atoms by a suitable reducing agent. Hence, most of the crystals with latent image nuclei are 'developed' to give, for the case of black-and-white film, a black dot of colloidal dimensions. This form of chemical response to a photon signal corresponds to an amplification of about 10^6.

The detailed chemistry of photographic emulsions is very involved and encompasses solid-state physics and surface science. In colour photography three different layers of emulsion are used where each layer is sensitized to one of the three primary colours (blue, green and red). Sensitizer molecules absorb light in each of these wavelength bands and transfer electrons to the surface of a nearby AgBr crystal. These molecules have to be within about 2 nm of an AgBr crystal to transfer the freed electron to a silver ion. During the development stage the reducing agent itself couples to form a dye after it has been oxidized. Surface and colloid science is of fundamental importance in producing the emulsion, coating the film and controlling the reaction between the dye–developer and the AgBr surface. In addition, finely divided colloidal TiO_2 is dispersed in printing paper to give a high level of light reflectance and brightness.

Table 5.1

Dispersed phase	Dispersion medium	Colloidal system	Example
gas	liquid	foam	froth and beer
liquid	liquid	emulsion	milk, mayonnaise
solid	liquid	liquid sol	jellies, starch solution
liquid	solid	solid sol	chocolate
solid	solid	solid sol	candy

Emulsions in food science

Colloid chemistry is very important in the production and storage of foods because of the common requirement that an organic nutrient compound must be dispersed in water (in which it may be insoluble) and this dispersion must be stabilised by surface forces. In addition, the texture or feeling of a food in the mouth depends critically on its colloidal size distribution. A categorization of colloidal systems used in foodstuffs is given in Table 5.1. Gum arabic is a 'hydrocolloid' (a hydrated polymer which increases viscosity) and is used to stabilize the foam on beer by reducing the rate of thinning of the soap films. The gas in the foam also seems to matter as evidenced by Guinness, whose fine cream foam is dependent on a mixture of carbon dioxide and nitrogen. Natural lipid surfactants such as the lecithins are used to stabilize oil-in-water and water-in-oil food emulsions and prevent the phases from separating out. It is for this reason that eggs (which contain lecithin) are used to mix oil and water in the production of sauces.

Industrial Report

Colloid science in foods

Competitive adsorption of molecules at interfaces, or the displacement of one stabilizing molecule by another, is an important process in food manufacture. Molecular rearrangement at interfaces not only affects the

Continued

processing of foods, but also the stability of the final product, the overall microstructure, and the sensory properties that the consumer ultimately experiences.

Ice cream is a complex colloidal dispersion consisting of ice particles, air bubbles, a semi-solid fat emulsion (or dispersion), protein aggregates, sugars, and polysaccharide viscosity modifiers. Ice cream is produced from a liquid mix which is emulsified (by homogenization) to form an oil-in-water emulsion (with droplet diameter of about 0.5 to 1 μm). The protein-stabilized emulsion is then rapidly cooled so that the oil starts to crystallize and become semi-solid particles. The cooled mix is then aerated and frozen simultaneously in a high shear process. The fat dispersion in the mix prior to freezing is under constant thermal motion, and its stability to flocculation and/or coalescence depends upon the forces that act between the particles. In general, the fat droplets are stable to coalescence due to the protein that adsorbs at the oil/water interface. This leads to stabilization through electrostatic and steric repulsion mechanisms.

Although the emulsion in the mix is relatively stable, ice cream manufacture actually requires a controlled amount of 'destabilization' of this emulsion. This is in order to enable fat particles to stick to the air bubbles and, to a controlled extent, one another, during the aeration and freezing step. Because the fat particles are 'semi-solid', they form partially coalesced fat particles as opposed to fully coalescing like liquid oil-in-water emulsions, i.e. two or more partially coalesced fat droplets will retain some of their structural integrity as opposed to forming one larger spherical droplet. This process leads to several important effects: increased stability of the air bubbles to growth, slower melting of the ice cream, and a more 'creamy', smooth, ice cream taste and texture.

Fat that is coated purely by dairy protein in the ice cream mix will remain stable even through the high shear forces of the freezer, i.e. little, or no, destabilization will result and no fat particles will adsorb to the air. This would lead to a faster-melting and less creamy ice cream product. In order to reduce the stability of the fat particles to shear in the freezing process, small-molecule surface-active agents are added to the ice cream mix prior to emulsification. These typically include monoglycerides of fatty acids, or Tween [commercial detergent]. On cooling of the emulsion, they compete with the protein for space at the oil–water interface. The smaller molecules displace some of the protein, leading to an interface that is less stable to the shear and collision processes in the freezer. Therefore, they are more likely to stick to air bubbles and remain there. It is possible to add too much of the destabilizing emulsifier, however. This would lead to an emulsion that is too unstable to shear. This can lead to the formation of very large fat particles through partial coalescence, a phenomenon known as 'buttering'. Therefore, the stability of the semi-solid oil-in-water dispersion needs to be carefully controlled in order to keep the product properties optimized.

Dr Andrew Cox
Unilever Research
Colworth
UK

Experiment 5.1   Determination of the phase behaviour of microemulsions

Introduction

Micelles are formed by most surfactants (especially single-chained ones) even at fairly low concentrations in water, whereas microemulsions can be produced at much higher surfactant concentrations with, of course, the addition of an oil (e.g. decane). Microemulsions are most readily formed by double-chained surfactants. The microemulsion region for a surfactant–oil–water mixture is determined and plotted on a triangular three-phase diagram. The microemulsion region is a single, clear phase because the aggregates are too small to significantly scatter light. The double-chained cationic surfactant used in this experiment is didodecyl-dimethylammonium bromide (DDAB) and decane is used as the oil. Three-phase diagrams of this type are often measured using an aqueous salt solution as the third phase but in this experiment we will use distilled water. An example of this type of triangular graph is given in Figure 5.4.

This type of graph has some interesting properties and must be used carefully. First of all, it should be noted that pure phases of the three components correspond to each apex of the triangle and that concentrations should be in either mole fractions or percentages. From this it is easy to see which concentration axis refers to any particular com-

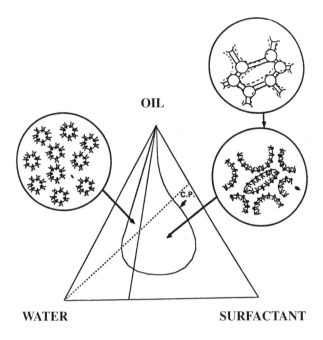

Figure 5.4 A typical three phase triangular diagram.

ponent. It is important to realize that the mole fraction noted on this axis relates to the line forming the base of the triangle where the apex corresponds to that of the pure phase (see diagram). Another important point is that a line drawn from any apex to the opposite axis (see diagram) corresponds to a constant ratio of mole fractions of the other two components. (In two-phase regions tie lines have to be drawn in order to denote the compositions of the two phases.)

Experimental details

The aim of the experiment is to determine the microemulsion (i.e. clear, single-phase) region for the three components already discussed and map out the results on a triangular phase diagram. The microemulsion region is determined by making up a series of mixtures in $10 \, cm^3$ stoppered Erlenmeyer flasks with compositions that span the anticipated range. The procedure is to start with a volume of about $2 \, cm^3$ of oil and surfactant mixtures spanning the range 20% decane to 80% decane by weight (i.e. at points along the oil-surfactant axis), weigh the sample and flask and then add increasing amounts of water up to about 70%, observing the properties of each mixture after thoroughly mixing, using a vibrator table. In this way the initial ratio in the mixture of oil to surfactant remains constant and each initial mixture runs along a different line to the pure water apex, as illustrated in the figure for the case of a starting ratio of 0.67 oil to 0.33 surfactant (by weight). At the point where a clear microemulsion phase is formed, the flask can be reweighed to obtain the amount of water added.

At low water contents (~10–20%) the mixtures will generally be milky and at some composition will become clear – at this composition a microemulsion is produced and the boundary point has been ascertained and can be plotted. On increasing the water content a second transition is reached (at typically about 60% water), which is more difficult to observe. This is the formation of a gel of high viscosity and marks the other boundary of the microemulsion region.

Using starting compositions of 10, 30, 50 and 70% decane add increasing amounts of water and hence map out the microemulsion region on a triangular graph similar to that shown here (but in terms of weight %). Note that the initial mixtures can be made more uniform (i.e. better mixed) by adding a small amount of water (no more than 2%). Also, after each addition of water the mixture should be gently shaken and then observed before adding more. At each stage the amount of water added must be known. Care must be taken to note down the visible properties of each mixture such as clarity and estimated viscosity. Also, the flasks should be carefully stoppered to prevent significant loss of decane.

Questions

(1)   What approximate composition of DDAB surfactant in water would you recommend to pump down an oil well to improve oil recovery if the main cost of the process was the cost of the surfactant?

(2)   How would you expect the structure of the aggregates in the microemulsion to vary as the % oil increases?

Experiment 5.2 Determination of the phase behaviour of concentrated surfactant solutions

Introduction

Micelles are spontaneously formed by most surfactants (especially single-chained ones) even at fairly low concentrations in water, whereas at higher surfactant concentrations, with or without the addition of an oil (e.g. octane) or co-surfactant (e.g. pentanol), a diverse range of structures can be formed. These various structures include micelles, multibilayers (liquid crystals), inverted micelles, emulsions (swollen micelles) and a range of microemulsions. In each case, the self-assembled structures are determined by the relative amounts of surfactant, hydrocarbon oil, co-surfactant (e.g. pentanol) and water, and the fundamental requirement that there be no molecular contact between hydrocarbon and water.

In this experiment we will use various experimental techniques to attempt to identify the structures formed by a range of oil, water, surfactant and co-surfactant mixtures. The clues to this identification will come from:

(1) the composition,

(2) visual observation,

(3) microscopic observations with normal and polarizing light,

(4) transmission of polarized light (using crossed polarizing films),

(5) viscosity, and

(6) electrical conductivity.

The structures typically formed are listed below:

- Micellar.

- Inverted micellar (emulsion or microemulsion).

- Lamellar (liquid crystalline).

- Swollen micellar (i.e. microemulsion).

- Bi-continuous microemulsion.

- Emulsions (oil-in-water and water-in-oil).

Experimental details

Use the techniques listed above (1–6) to assign the structures formed by the following mixtures. You may wish to consult a demonstrator for help with identification techniques. In your report, explain the reasons for each of your structural assignments.

Make two identical surfactant solutions by dissolving 2.5 g sodium dodecyl sulphate (SDS) in 10.0 cm³ of distilled water at room temperature. (The solution may need to be heated to dissolve the surfactant.) Take care: excessive shaking will cause foaming.

To one of these solutions add pentanol sequentially, swirling to mix and observe after each mixture has equilibrated (about 5 min).

Series I

Sample No.	Amount added/(g)	Total amount added/(g)
1	0	0
2	0.100	0.100
3	0.900	1.000
4*	2.250	3.250
5	5.250	8.500
6	10.250	18.750

(*Note: you will not be able to measure the conductivity of this sample).

Present the results in your report in the form of a table, listing the % composition (by weight) of each sample and the corresponding observations, and suggested structures.

After the final addition of alcohol to Series I, add 20.0 g of octane and examine the resulting mixture.

Series II

Repeat the above procedure on the second SDS solution with octane rather than pentanol and again construct a table of your observations and conclusions. (Note that samples 3–6 in this case are unsuitable for conductivity measurements.)

Source

This experiment is adapted from Friberg and Beniksen (1979).

Question

Explain why light can be transmitted through crossed polarizing films when a birefringent sample is placed between them.

6

Charged Colloids

The generation of colloidal charges in water. The theory of the diffuse electrical double-layer. The zeta potential. The flocculation of charged colloids. The interaction between two charged surfaces in water. Laboratory project on the use of microelectrophoresis to measure the zeta potential of a colloid.

The formation of charged colloids in water

Most solids release ions to some extent when immersed in a high-dielectric-constant liquid such as water. Even oil droplets and air bubbles have a significant charge in water. In order for this charging process to take place ions must dissociate from the surface, which will then be of opposite charge. (Note that the charge could equally well be created by the selective adsorption of a particular ion from solution.) Let us consider the work required to separate an ion pair in vacuum (or air) and in a high-dielectric-constant liquid, such as water. This situation is illustrated in Figure 6.1. D is the static dielectric constant of the medium, which is given by the ratio of permittivities of the medium (ε) and of free space (ε_0). From Coulomb's law for two charges, q_1 and q_2, separated by distance r, the attractive force F_c is given by the relation:

Applied Colloid and Surface Chemistry Richard M. Pashley and Marilyn E. Karaman
© 2004 John Wiley & Sons, Ltd.   ISBN 0 470 86882 1 (HB)   0 470 86883 X (PB)

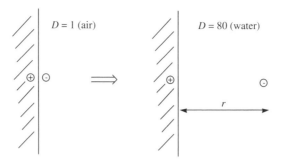

Figure 6.1   Diagram illustrating the ionization of a surface in air and water.

$$F_c = \frac{q_1 q_2}{4\pi D\varepsilon_0 r^2} \qquad (6.1)$$

where the force F_c is attractive (negative) between unlike charges. As a first approach, we can easily estimate the work (W_c) required to separate two charges from close separation of, say, $2\,\text{Å}$ to a large distance by integration of (6.1). In water, where $D = 80$, this work is about $6 \times 10^{-21}\,\text{J}$, which is quite close to the kinetic energy of the free ion (i.e. kT). However, if the medium was air or vacuum ($D = 1$), the work required would be about $100\,kT$. Clearly, it is the high dielectric constant or polar nature of water which allows this ion dissociation to occur, whereas in air and non-polar liquids (e.g. $D \sim 2$ for liquid hexane) we would expect no dissociation. It is for this reason that we are mostly interested in charging processes at the solid/aqueous interface and the stability of colloids, which often become charged when dispersed in aqueous media.

The theory of the diffuse electrical double-layer

Although the single ion dissociation approach gives a good indication of the basic conditions for dissociation, the real situation is more complicated because there will usually be a high density of ions dissociated from the surface. This will introduce repulsive forces between the dissociated ions and a much stronger attraction back to the surface, because of the high electric field generated there. In fact, the electric field generated at the surface prevents the dissociated ions from leaving the surface region completely and these ions, together with the charged

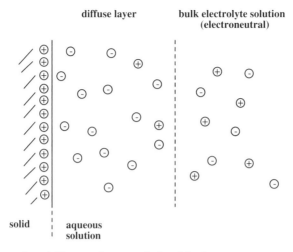

Figure 6.2 The diffuse electrical double-layer in aqueous solution next to a flat surface.

surface, form a 'diffuse electrical double-layer', as illustrated in Figure 6.2.

In order to understand the diffuse layer in detail, we need to go back to the fundamental equations of electrostatics due to J.C. Maxwell. The equation of interest relates the local electric field $\vec{E}(\vec{r})$ at the position vector $\vec{r}$ to the net local electric charge density $\rho(\vec{r})$:

$$\vec{\nabla} \bullet \vec{E}(\vec{r}) = \frac{\rho(\vec{r})}{\varepsilon_0 D} \tag{6.2}$$

Thus, the divergence or flux of the electric field is directly related to the net charge at that point. The electric field is simply defined as the force acting on a unit charge:

$$\vec{E}(\vec{r}) = \lim_{q \to 0} \frac{\vec{F}(\vec{r})}{q} \tag{6.3}$$

The limit is required because otherwise a finite test charge 'q' would itself perturb the electric field. Clearly, from (6.3), the direction of the electric field is that taken by a positive charge when free to move. Because the electric field is a vector, it is often more useful to use the corresponding electrostatic potential $\psi(\vec{r})$, which is a scalar quantity and is defined as the potential energy gained by moving a unit charge from infinity to the position $\vec{r}$. The potential energy of an ion of charge

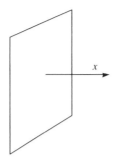

Figure 6.3 The one-dimensional case of a flat surface.

$Z_i q$ at $\vec{r}$ is therefore simply given by $Z_i q\psi(\vec{r})$, where q is the proton charge and Z_i the valency of the ion. Note that if the potential is positive, then a positive charge will have an increased (positive) potential energy compared with that at infinity.

Since $\psi(\vec{r})$ is the electrostatic potential energy per unit charge, the gradient of this parameter with distance must be equal to the force acting on a unit charge – which is the definition of the electric field. Hence it follows that

$$\vec{E}(\vec{r}) = -\nabla\psi(\vec{r}) \tag{6.4}$$

In order to simplify (6.2), let us consider the common case of the electric field generated by a charged, flat surface (such as an electrode), see Figure 6.3. For this case and using electric potentials (6.2) becomes

$$\frac{d^2\psi(x)}{dx^2} = -\frac{\rho(x)}{\varepsilon_0 D} \tag{6.5}$$

where ψ and ρ are now only functions of x, the distance from the flat surface. To solve this equation we need to find the relationship between $\rho(x)$ and $\psi(x)$. The local density of any ion of charge $Z_i q$ (which can be either positive or negative) must depend on its electrostatic potential energy at that position (i.e. at x). From our definition of ψ this potential energy is given by $Z_i q\psi(x)$. Since any ion next to a charged surface must be in equilibrium with the corresponding ions in the bulk solution, it follows that the electrochemical potential of an ion at distance x from the surface must be equal to its bulk value. Thus:

$$\mu_i^b = \mu_i^x = \mu_i^0 + kT \ln C_i(B) = \mu_i^0 + Z_i q\psi(x) + kT \ln C_i(x)$$

where $C_i(B)$ and $C_i(x)$ are the ion concentrations in bulk and at distance x from the charged surface, and it is assumed that these are dilute solutions (i.e. $\psi(B) = 0$). This equation leads directly to the Boltzmann distribution, which can be used to obtain the concentration at any other electrostatic potential energy by the familiar relationship

$$C_i(x) = C_i(B)\exp\left[-\frac{Z_i q \psi(x)}{kT}\right] \tag{6.6}$$

This result is very useful because it gives us the concentration of any ion next to a charged surface when immersed in an electrolyte solution.

We can also use this result to obtain the net charge density (due to both counterions and co-ions) at a distance x from a charged surface:

$$\rho(x) = \sum_i Z_i q C_i(x)$$
$$= \sum_i Z_i q C_i(B)\exp\left[\frac{-q\psi(x)Z_i}{kT}\right] \tag{6.7}$$

and substitution in (6.5) then gives the result

$$\frac{d^2\psi(x)}{dx^2} = -\frac{q}{\varepsilon_0 D}\sum_i Z_i C_i(B)\exp\left[\frac{-Z_i q \psi(x)}{kT}\right] \tag{6.8}$$

This is the important 'Poisson–Boltzmann' (PB) equation and the model used to derive it is usually called the 'Gouy–Chapman' (GC) theory. It is the basic equation for calculating all electrical double-layer problems, for flat surfaces. In deriving it we have, however, assumed that all ions are point charges and that the potentials at each plane 'x' are uniformly smeared out along that plane. These are usually reasonable assumptions.

We can now set about using (6.8) to give us information about the quantitative details of the electrical double-layer. Let us, for simplicity, assume that the electrolyte in which the surface is immersed is symmetrical, that is a $Z:Z$ electrolyte (i.e. 1:1, 2:2 or 3:3, where $Z = 1$, 2 or 3). Equation (6.7) then becomes

$$\rho(x) = ZqC(B)\left\{\exp\left[\frac{-Zq\psi(x)}{kT}\right] - \exp\left[\frac{Zq\psi(x)}{kT}\right]\right\}$$

and if we, for convenience, define $Y = Zq\psi(x)/kT$ this reduces to

$$\rho(x) = -ZqC(B)\{\exp(Y) - \exp(-Y)\}$$

$$\therefore \rho(x) = -2ZqC(B)\sinh(Y) \tag{6.9}$$

which in the PB equation (6.8) gives

$$\frac{d^2\psi(x)}{dx^2} = \frac{2Z^2q^2C(B)}{\varepsilon_0 DkT}\sinh(Y) \tag{6.10}$$

(Note that for symmetrical electrolytes $C(B) = C_i(B)$.)

Now, for convenience (and not arbitrarily, as will be seen later), let us replace the real distance x with a scaled distance X, such that $X = \kappa x$ and κ^{-1}, which is called the *Debye length*, is defined as:

$$\kappa^{-1} = \left\{\frac{\varepsilon_0 DkT}{q^2 \sum_i C_i(B)Z_i^2}\right\}^{1/2} \tag{6.11}$$

The useful parameter κ^{-1} has the units of length and depends on both the electrolyte type and concentration. Scaling x in (6.10) thus gives the deceptively simple, non-linear, second-order differential equation

$$\frac{d^2Y}{dX^2} = \sinh Y \tag{6.12}$$

Integration of (6.12) gives the potential distribution next to a charged surface:

$$Y = 2\ln\left[\frac{1 + \gamma\exp(-X)}{1 - \gamma\exp(-X)}\right] \tag{6.13}$$

where

$$\gamma = \left[\frac{\exp(Y_0/2) - 1}{\exp(Y_0/2) + 1}\right]$$

and Y_0 is the scaled electrostatic potential at the surface of the charged plane (i.e. at $x = 0$, $\psi = \psi_0$ and at $X = 0$, $Y = Y_0$). This result gives us a great deal of information about the decay in potential and hence the distribution of all ionic species in the diffuse electrical double-layer next to a charged, flat surface.

The Debye length

We can easily get some idea of what this theory predicts by looking at the limit of low potentials, that is where $Y_0 \ll 1$ (which for $1:1$ electrolytes corresponds to $\psi_0 < 25\,\text{mV}$). In this case (6.13) can be shown to reduce to the simple result

$$\psi(x) \cong \psi_0 \exp(-\kappa x) \tag{6.14}$$

which demonstrates the physical meaning of the Debye length. This length is also referred to as the 'double-layer thickness' and, from (6.14), it obviously gives an indication of the extent of the diffuse layer. It is the distance from the surface where the surface potential has fallen to 1/e of its original value. Equation (6.14), although approximate, enables us to estimate the decay in electrostatic potential away from a flat, charged surface. Some typical results are shown in Figure 6.4. These results clearly demonstrate the effect of electrolyte concentration

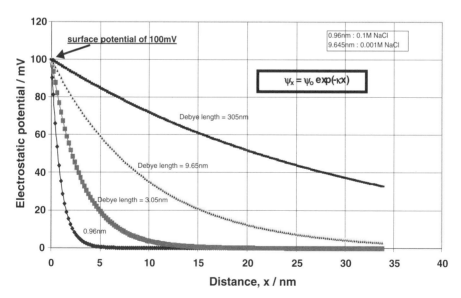

Figure 6.4 Estimates of the decay in electrostatic potential away from a charged flat plate in a range of electrolyte solutions.

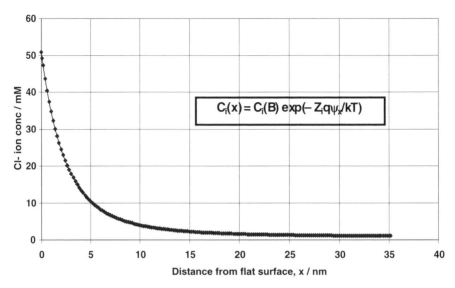

Figure 6.5 Estimates of the Ce⁻ counter-ion concentrations away from a flat, charged surface of potential +100 mV.

on screening the range of the surface electrostatic potential. In each case the Debye length is equal to the distance from the surface at which the surface potential has fallen to 100/e or 37 mV. (Note that use of (6.13) gives a more accurate potential distribution.)

Once the potential distribution next to a charged surface is known the Boltzmann distribution (6.6) can be used to calculate the corresponding distribution of both counter-ions and co-ions, in an electrolyte solution next next to a charged surface. The results obtained for both counter-ions and co-ions are given in Figures 6.5 and 6.6 for a monovalent electrolyte solution, e.g. NaCl electrolyte at mM concentration next to a +100 mV flat surface. Using the relatively simple equations derived above it is clearly possible to obtain precise details about the ion distributions next to charged, flat surfaces when they are immersed in electrolyte solutions. This information is crucial to a thorough understanding of the properties of electrodes, colloidal solutions and soil chemistry. As might be expected, counter-ions are pulled towards surfaces of opposite charge, reaching quite high concentrations close to the surface. By comparison, co-ions are expelled by the same-charge surface and form a depleted layer. Similar results can be obtained using a combination of mixed and multi-valent electrolytes.

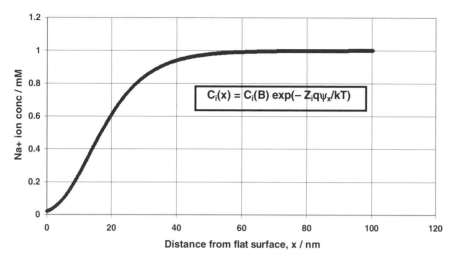

Figure 6.6 Estimates of the Na$^+$ co-ion concentrations away from a flat, charged surface of potential +100 mV.

The surface charge density

So far we have talked only in terms of electrostatic potentials. Can we use this theory to find the charge density on the surface (σ_0)? In order for the electrical double-layer to be neutral overall, it follows that the total summed charge in the diffuse layer must equal the surface charge. Thus, it follows that

$$\sigma_D = -\sigma_0 = \int_0^\infty \rho(x)\,dx \qquad (6.15)$$

where σ_D is the total diffuse layer charge. Now, since from (6.5),

$$\frac{d^2\psi(x)}{dx^2} = \frac{-\rho(x)}{\varepsilon_0 D} \qquad (6.5)$$

it follows that

$$\sigma_D = \varepsilon_0 D \int_0^\infty \frac{d^2\psi(x)}{dx^2}\,dx$$

and hence

$$\sigma_0 = -\varepsilon_0 D \left[\frac{d\psi(x)}{dx} \right]_{x=0} \tag{6.16}$$

since $d\psi(x)/dx = 0$ as $x \rightarrow \infty$. Now we can integrate (6.8) using the fact that

$$\int \frac{d^2 y}{dx^2} dy = 1/2 \left(\frac{dy}{dx} \right)^2 + C \tag{6.17}$$

and using various known boundary conditions obtain the first-order differential equation

$$\frac{d\psi(x)}{dx} = -\left\{ \frac{2kT}{\varepsilon_0 D} \sum_i C_i(B) \left[\exp \left(-\frac{Z_i q \psi(x)}{kT} \right) - 1 \right] \right\}^{1/2} \tag{6.18}$$

which on substitution into (6.16) gives the result

$$\sigma_0 = (\operatorname{sgn} \psi_0) \left\{ 2\varepsilon_0 DkT \sum_i C_i(B) \left[\exp \left(-\frac{Z_i q \psi_o}{kT} \right) - 1 \right] \right\}^{1/2} \tag{6.19}$$

which, for a 1:1 electrolyte reduces to

$$\sigma_0 = [8\varepsilon_0 DkTC(B)]^{1/2} \sinh \left(\frac{q\psi_o}{2kT} \right) \tag{6.20}$$

Hence, the surface charge can be easily calculated from the surface potential for a planar surface.

The zeta potential

The stability of many colloidal solutions depends critically on the magnitude of the electrostatic potential (ψ_0) at the surface of the colloidal particles. One of the most important tasks in colloid science is therefore to obtain an estimate of ψ_0 under a wide range of electrolyte conditions. In practice, one of the most convenient methods for obtaining ψ_0 uses the fact that a charged particle will move at some constant, limiting velocity under the influence of an applied electric field. Even quite small particles (i.e. <1 μm) can be observed using a dark-field microscope and their velocity directly measured. This technique is called *microelectrophoresis* and what is measured is the electromobility (μ) of a colloid, that is its speed (u) divided by the applied electric field (E).

Let us now examine how we can obtain an estimate of ψ_0 from the measured electromobility of a colloidal particle. It turns out that we can obtain simple, analytic equations only for the cases of very large and very small particles. Thus, if a is the radius of an assumed spherical colloidal particle, then we can obtain direct relationships between electromobility and the surface potential, if either $\kappa a > 100$ or $\kappa a < 0.1$, where κ^{-1} is the Debye length of the electrolyte solution. Let us first look at the case of small spheres (where $\kappa a < 0.1$), which leads to the Hückel equation.

The Hückel equation ($\kappa a < 0.1$)

The spherically symmetric potential around a charged sphere is described by the Poisson–Boltzmann equation:

$$\frac{1}{r^2} \frac{d}{dr}\left[r^2 \frac{d\psi}{dr} \right] = -\frac{\rho(r)}{\varepsilon_0 D} \tag{6.21}$$

where $\rho(r)$ is the charge density and ψ the potential at a distance r away from a central charge. This equation can be simplified using the Debye–Hückel or linear approximation valid for low potentials:

$$\frac{1}{r^2} \frac{d}{dr}\left[r^2 \frac{d\psi}{dr} \right] = \kappa^2 \psi \tag{6.22}$$

which has the simple, general solution:

$$\psi = \frac{A \exp(\kappa r)}{r} + \frac{B \exp(-\kappa r)}{r} \tag{6.23}$$

The constant A must equal zero for the potential ψ to fall to zero at a large distance away from the charge and the constant B can be obtained using the second boundary condition, that $\psi = \psi_0$ at $r = a$, where a is the radius of the charged particle and ψ_0 the electrostatic potential on the particle surface. Thus we obtain the result that

$$\psi_0 = \frac{B \exp(-\kappa r)}{r} \qquad (6.24)$$

and, therefore

$$\psi = \frac{\psi_0 a \exp[-\kappa(r-a)]}{r} \qquad (6.25)$$

The relationship between the total charge q on the particle and the surface potential is obtained using the fact that the total charge in the electrical double-layer around the particle must be equal to and of opposite sign to the particle charge, that is:

$$q = -\int_a^\infty 4\pi r^2 \rho(r) dr \qquad (6.26)$$

where $\rho(r)$ is the charge density at a distance r from the centre of the charged particle. The value of $\rho(r)$ can be obtained from combination of (6.21) and (6.22), assuming the linear approximation is valid and hence

$$q = 4\pi\varepsilon_0 D\kappa^2 \int_a^\infty r^2 \psi dr \qquad (6.27)$$

Now, using (6.25) for ψ,

$$q = 4\pi\varepsilon_0 D\kappa^2 a\psi_0 \int_a^\infty r \exp[-\kappa(r-a)] dr \qquad (6.28)$$

Integration using Leibnitz's theorem gives:

$$q = 4\pi\varepsilon_0 Da\psi_0(1 + \kappa a) \qquad (6.29)$$

and rearranging this equation leads to a useful physical picture of the potential around a sphere, thus:

$$\psi_0 = \frac{q}{4\pi D\varepsilon_0 a} - \frac{q}{4\pi D\varepsilon_0(a + \kappa^{-1})} \qquad (6.30)$$

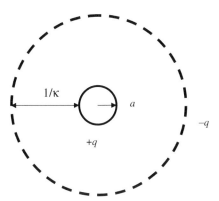

Figure 6.7 Diagram of the diffuse electrical double-layer around a small, charged colloid.

This result corresponds to a model of the charged particle with a diffuse layer charge (of opposite sign) at a separation of $1/\kappa$, as illustrated in Figure 6.7.

Since we now have (6.29) which relates the charge on the particle to the surface potential, we can combine this with the forces acting on a moving particle in an applied electric field. Thus, when the particle is moving at constant velocity (u) the electrostatic force on the particle (qE) must equal the drag force, which we may assume (for laminar, steady fluid flow) to be that given by Stokes's law (i.e. $F_{drag} = 6\pi a u \eta$). Using (6.29) and the fact that we define the electromobility (μ) of a particle as u/E, we obtain the result that

$$\psi_0 = \frac{3\mu\eta}{2\varepsilon_0 D(1+\kappa a)} \tag{6.31}$$

which for $\kappa a \ll 1$ becomes:

$$\psi_0 = \frac{3\mu\eta}{2\varepsilon_0 D} = \zeta \tag{6.32}$$

In this result, the condition of small particles means that the actual size of the particles (which is often difficult to obtain) is not required. For reasons to be discussed later, we will call the potential obtained by this method the *zeta potential* (ζ) rather than the surface potential. In the following section we consider the alternative case of large colloidal particles, which leads to the Smoluchowski equation.

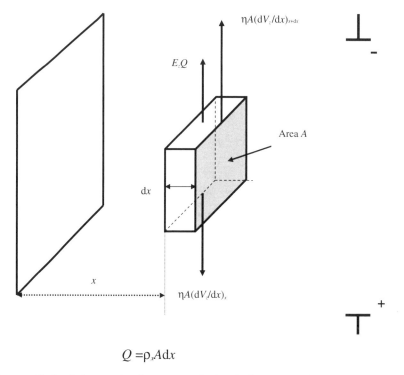

$$Q = \rho_x A dx$$

Figure 6.8 Schematic diagram of the balance in forces acting on a fluid element close to the surface of a large colloidal particle.

The Smoluchowski equation ($\kappa a > 100$)

Let us now consider an alternative derivation for the case of large colloidal particles, where the particle radius is much larger than the Debye length (i.e. $\kappa a > 100$). The situation is best described by the schematic diagram, Figure 6.8, where the surface of the large particle is assumed to be effectively flat, relative to the double-layer thickness. It is also assumed, in this approach, that the fluid flows past the surface of the particle in parallel layers of increasing velocity with distance from the surface. At the surface the fluid has zero velocity (relative to the particle) and at a large distance away, the fluid moves with the same velocity as the particle, but in the opposite direction. It is also assumed that the flow of the fluid does not alter the ion distribution in the diffuse double-layer (i.e. in the x direction). Under these conditions mechanical equilibrium can be considered in a fluid element between x and $x + dx$, when the viscous forces acting in the z direction on the fluid element, due to the velocity gradient in the x direction, are precisely balanced by the electrostatic body force acting on the fluid due the charge contained in it. Thus, we obtain the mechanical equilibrium condition that

$$E_z \rho_x A dx = \eta A \left(\frac{dV_z}{dx} \right)_x - \eta A \left(\frac{dV_z}{dx} \right)_{x+dz} \tag{6.33}$$

or

$$E_z \rho_x A dx = -\eta A \left(\frac{d^2 V_z}{dx^2} \right) dx \tag{6.34}$$

We can then relate the charge density, ρ_x, to the electrostatic potential using the one-dimensional Poisson–Boltzmann equation,

$$\frac{d^2 \psi}{dx^2} = -\frac{\rho_x}{\varepsilon_0 D} \tag{6.35}$$

Thus, in (6.34):

$$E_z \varepsilon_0 D \frac{d^2 \psi}{dx^2} dx = \eta \left(\frac{d^2 V_z}{dx^2} \right) dx \tag{6.36}$$

which on integration gives

$$E_z \varepsilon_0 D \frac{d\psi}{dx} = \eta \left(\frac{dV_z}{dx} \right) + c_1 \tag{6.37}$$

Since $d\psi/dx = 0$ when $dV_z/dx = 0$, the integration constant, c_1, must be equal to zero and a second integration,

$$\int_{\psi=0}^{\psi=\varsigma} E_z \varepsilon_0 D \frac{d\psi}{dx} dx = \int_{V_z}^{0} \eta \left(\frac{dV_z}{dx} \right) dx \tag{6.38}$$

produces the result that

$$E_z \varepsilon_0 D \varsigma = -\eta V_z \tag{6.39}$$

if it is assumed that $D \neq f(x)$ and $\eta \neq f(x)$ (i.e. that the fluid is Newtonian). Since $-V_z$ refers to the fluid velocity we can easily convert

(6.39) to particle velocity (i.e. $V_p = -V_z$) and from our definition of electromobility (μ) it follows that

$$\varsigma = \frac{\mu\eta}{\varepsilon_0 D} \tag{6.40}$$

This important result is called the *Smoluchowski equation* and, as before, the zeta potential is directly related to the mobility and does not depend on either the size of the particle or the electrolyte concentration.

In summary, for the two extreme cases, we have:

$$\varsigma = \frac{3\mu\eta}{2\varepsilon_0 D}, \quad \text{for } \kappa a \ll 1\,(<0.1)$$

$$\varsigma = \frac{\mu\eta}{\varepsilon_0 D}, \quad \text{for } \kappa a \gg 1\,(>100)$$

Corrections to the Smoluchowski equation

These equations have been shown to be correct under the conditions of electrolyte concentration and particle size stated. However, it is easy to show that using typical colloidal sizes and salt concentrations, many colloidal systems of interest, unfortunately, will fall between the ranges covered by these equations. In deriving both the Hückel and Smoluchowski equations we have, for simplicity, ignored 'relaxation' and 'electrophoretic retardation' effects, which have to be included in a more complete theory. The origin of these effects is illustrated in Figure 6.9, which shows a solid, charged particle moving in an electric field.

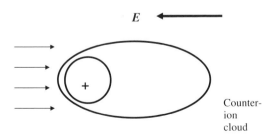

Figure 6.9 Diagram of a charged colloid moving in a fluid under the action of an electric field.

When in motion, the diffuse electrical double-layer around the particle is no longer symmetrical and this causes a reduction in the speed of the particle compared with that of an imaginary charged particle with no double-layer. This reduction in speed is caused by both the electric dipole field set up which acts in opposition to the applied field (the relaxation effect) and an increased viscous drag due to the motion of the ions in the double-layer which drag liquid with them (the electrophoretic retardation effect). The resulting combination of electrostatic and hydrodynamic forces leads to rather complicated equations which, until recently, could only be solved approximately. In 1978, White and O'Brien developed a clever method of numerical solution and obtained detailed curves over the full range of κa values ($0 \rightarrow \infty$) and surface potentials. At intermediate values of κa the relationship between ζ and μ is non-linear and strongly dependent on electrolyte type (i.e. charge and ion diffusion coefficients). A computer program is available from these authors to enable calculation of the zeta potential for any common electrolyte. With the successful introduction of this precise numerical procedure the onus is now on experimentalists to carry out well-defined mobility measurements on ideal, spherical particles of accurately known size.

Figure 6.10 shows the non-dimensional electrophoretic mobility as a function of zeta potential, for a range of κa values, corrected for these two factors. The curves represent computed values obtained by White and O'Brien, while the broken lines are the thin double-layer approximation. The $\kappa a = \infty$ line is Smoluchowski's result. (See also the standard text *Zeta Potential in Colloid Science* by R.J. Hunter, 1981.)

Finally, let us return to the problem of relating the measured zeta potential to the defined surface potential. The zeta potential is always measured (by definition) in an electrokinetic experiment. In this case the fluid has to flow around the particle. We expect, however, that a certain thickness of fluid (of roughly molecular dimensions) will remain stationary with respect to the particle, due to the large amount of work required to move fluid molecules along a solid surface. Obviously there will not be a sharp cut-off at say one or two molecular layers but a gradual increase in fluid flow will occur away from the particle. Since the total charge on the solid particle is responsible for the surface potential, the measured value, zeta, is generally of slightly lower magnitude. As we will see later, a better estimate of the surface potential can be obtained from direct interaction force measurements, and values so obtained can be compared with electrokinetic measurements, on exactly the same system in some cases (such as for muscovite mica).

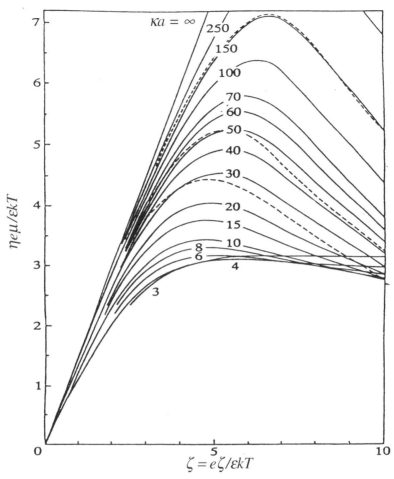

Figure 6.10 Theoretical calculations of the corrections required for the zeta potention (ζ) from electromobility (μ) values.

Some systems do show excellent agreement between zeta and the surface potential, whilst others differ significantly. Since no definitive results have yet been obtained, it is perhaps best to assume as a first approximation that these two potentials are similar.

The zeta potential and flocculation

A good example of the use of microelectrophoresis experiments is supplied by the study of ferric flocs, which are widely used in municipal water treatment plants. The zeta potentials shown below were derived from the measured floc electromobilities using the Smoluchowski equa-

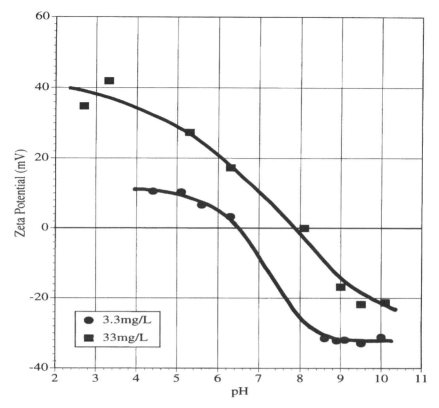

Figure 6.11 The zeta potential of ferric flocs as a function of concentration and PH.

tion. Figure 6.11 shows the effect of pH on the zeta potential of ferric flocs generated by precipitation of 3.3 mg/L (20 μM) and 33 mg/L (200 μM) ferric chloride in mM sodium chloride solution. Since almost all colloidal contaminants in natural water systems are negatively charged, it is important to operate under pH conditions where the flocs are positively charged, that is at pH values above the isoelectric point (iep, where the particles are uncharged). Under these conditions, the flocs have an electrostatic force attracting the contaminant particles to the flocs. The large flocs, containing adsorbed contaminants, are then sedimented and collected via filtration through high-flow-rate sand-bed filters. Unfortunately, in practice, ferric chloride usually contains manganese as a contaminant and this necessitates precipitation at pH values between 8 and 9, to prevent manganese dissolution, which will colour the drinking water. The use of these higher pH values means that polycationic, water-soluble polymers are often added, which adsorb on to the ferric flocs and increase their iep.

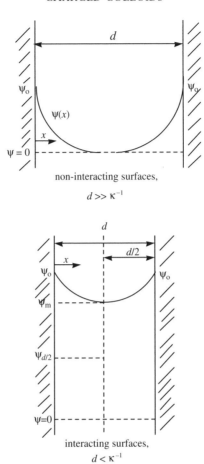

Figure 6.12 Schematic diagram of non-interacting and interacting charged surfaces.

The interaction between double-layers

So far, we have used the Maxwell equations of electrostatics to determine the distribution of ions in solution around an isolated, charged, flat surface. This distribution must be the equilibrium one. Hence, when a second surface, also similarly charged, is brought close, the two surfaces will 'see' each other as soon as their diffuse double-layers overlap. The ion densities around each surface will then be altered from their equilibrium value and this will lead to an increase in energy and a repulsive force between the surfaces. This situation is illustrated schematically in Figure 6.12 for non-interacting and interacting flat surfaces.

These repulsive electrostatic forces between similarly charged particles will act to prevent coagulation and stabilize colloidal solutions. If this repulsion were absent (for example, by neutralization of the surface charge), attractive van der Waals forces (to be discussed later) would cause each particle collision to be 'successful'. That is, particles colliding due to random kinetic motion in the solution would adhere to each other, forming large aggregates which settle out from solution. The system would coagulate. We can now apply the Poisson–Boltzmann model to this important interaction.

If we consider the interaction of two identical planar surfaces (see Figure 6.12), we see that at the mid-plane there is no net electric field, that is at $x = d/2$, $d\psi(x)/dx = 0$ and $\psi(x) = \psi_m$. Thus, even though there is a net (non-zero) charge on the fluid at the mid-plane, no net electrostatic force acts upon it. Now, since the mid-plane potential is not zero for interacting surfaces, there must be a higher concentration of ions at this plane compared with that in the bulk solution, from the Boltzmann distribution. This higher concentration, at the mid-plane, will give rise to a higher osmotic pressure at this plane, relative to the bulk solution, which will push the surfaces apart by drawing in more solvent (water). Since this pressure must be the same throughout the liquid film (between the surfaces) and because only osmotic forces act at the mid-plane, the pressure acting between the surfaces must be equal to the increased osmotic pressure at the mid-plane, over that in bulk. This osmotic pressure difference can be calculated simply by determining the total concentration of ions at the mid-plane (C_m^T). This concentration can be easily calculated using the Boltzmann distribution equation, if the mid-plane potential, ψ_m, is known, thus:

$$C_m^T = \sum_i C_i(B)\exp\left[-\frac{Z_i q \psi_m}{kT}\right] \tag{6.41}$$

which for (symmetrical) $Z:Z$ electrolytes becomes

$$C_m^T = C(B)\left\{\exp\left[-\frac{Zq\psi_m}{kT}\right] + \exp\left[\frac{Zq\psi_m}{kT}\right]\right\} \tag{6.42}$$

$$= 2C(B)\cosh\left[\frac{Zq\psi_m}{kT}\right] \tag{6.43}$$

For ideal solutions the osmotic pressure is simply given by CkT. Hence, the osmotic pressure difference between the mid-plane region and the bulk solution, which must be equal to the repulsive pressure in the film Π_R, is given by the equation

$$\Delta\Pi_{os} = \Pi_R = kT[C_m^T - 2C(B)]$$

and hence

$$\Pi_R = 2C(B)kT\left[\cosh\left(\frac{Zq\psi_m}{kT}\right) - 1\right] \qquad (6.44)$$

If we can now determine ψ_m as a function of the separation distance d between the surfaces, we can calculate the total double-layer (pressure) interaction between the planar surfaces. Unfortunately, the PB equation cannot be solved analytically to give this result and instead numerical methods have to be used. Several approximate analytical equations can, however, be derived and these can be quite useful when the particular limitations chosen can be applied to the real situation.

One of the simplest equations is obtained using the Debye–Hückel approximation (for low potentials) and the superposition principle. The latter assumes that the unperturbed potential near a charged surface can be simply added to that potential due to the other (unperturbed) surface. Thus, for the example shown in the Figure 6.12, it follows that $\psi_m = 2\psi_{d/2}$. This is precisely valid for Coulomb-type interactions, where the potential at any point can be calculated from the potentials produced by each fixed charge, individually. However, the Poisson–Boltzmann equation is non-linear (this has to do with the fact that in the diffuse double-layer the ions are not fixed but move because of their kinetic energy) and so this is formally not correct although it still offers a useful approximation.

Using the Debye–Hückel (DH) approximation the potential decay away from each flat surface is given by

$$\psi(x) \approx \psi_0 \exp(-\kappa x)$$

and hence, using the superposition principle,

$$\psi_m \approx 2\psi_{d/2} \approx 2\psi_0 \exp(-\kappa d/2) \qquad (6.45)$$

Again, if we use the DH approximation in (6.44), we can expand the cosh function for small values of x (or ψ_0) to give

$$\cosh x = (1 + x^2/2) \text{ or } \cosh(Zq\psi_m/kT) = 1 + 1/2(Zq\psi_m/kT)^2$$

Which simplifies (6.44) to

$$\Pi_R \approx C(B)kT\left(\frac{Zq\psi_m}{kT}\right)^2 \tag{6.46}$$

Hence, from (6.45) and (6.46),

$$\Pi_R \approx \frac{C(B)Z^2q^2}{kT}4\psi_0^2\exp(-\kappa d)$$

which on using the definition of the Debye length (κ^{-1}) becomes

$$\Pi_R \approx 2\varepsilon_0 D\kappa^2\psi_0^2\exp(-\kappa d) \tag{6.47}$$

This result shows us that the repulsive double-layer pressure (for the case of low potentials) decays exponentially with a decay length equal to the Debye length and has a magnitude which depends strongly on the surface potential.

The corresponding interaction energy V_F between flat surfaces can be obtained by integrating the pressure from large separations down to separation distance, d:

$$V_F = -\int_0^\infty \Pi_R dd \tag{6.48}$$

which gives

$$V_F \approx 2\varepsilon_0 D\kappa\psi_0^2\exp(-\kappa d) \tag{6.49}$$

So far, we have only considered the interaction between flat surfaces, basically because of the simplification of the PB equation in one dimension. Of course, colloidal particles are usually spherical and for this geometry the exact numerical solution of the three-dimensional PB equation becomes very difficult. However, we can obtain an estimate of the sphere–sphere interaction from the planar result if the radius a of the spheres is much larger than the Debye length (i.e. $\kappa a \gg 1$). This method was developed by Derjaguin.

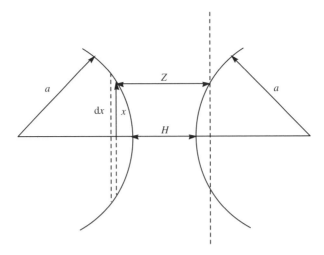

Figure 6.13 Diagram used to explain the Derjaguin approximation for the interaction between two spheres.

The Derjaguin approximation

A schematic diagram of the analysis method used is given in Figure 6.13. In this procedure, called the 'Derjaguin approximation', we consider the interaction of the circular annulus (dx) with an imaginary parallel surface plane at distance Z. With this assumption, the total interaction energy V_s between the spheres is then given by

$$V_S \approx 2\pi \int_{Z=H}^{Z=\infty} V_F x dx \tag{6.50}$$

From simple geometry we can easily rearrange this equation in the form

$$V_S \approx \pi a \int_{Z=H}^{Z=\infty} V_F dZ \tag{6.51}$$

Using the result given in (6.49), we can then obtain the corresponding interaction energy between spheres:

$$V_S \approx 2\pi a \varepsilon_0 D \psi_0^2 \exp(-\kappa H) \tag{6.52}$$

Once again the interaction energy decays exponentially and is strongly dependent on both the surface potential and the electrolyte concentration.

Industrial Report

The use of emulsions in coatings

From 2000 BC, when decorative paintings adorned the walls of Egyptian tombs, all the way up through much of the 1900s, little changed in the rudimentary approach to formulating decorative or protective paints. Paints were based on naturally occurring oleoresinous materials: combinations of naturally occurring drying oils and resins. Linseed oil became the most widely used oil, while amber was the most common resin. Eventually, naturally occurring resins were replaced by synthetic resins, such as alkyds. The resins and drying oils perform the role of the 'binder' in the paint; that is, they bind the pigment particles and stick to the substrate being painted – they are the 'glue'. These types of paints were usually thinned with organic solvents, consequently they are flammable and often represent a health hazard. The high levels of organic solvents present in these coatings causes considerable pollution concerns.

The architectural coatings industry was revolutionized around 1950, when coatings based on waterborne emulsions were invented. This paradigm shift represents the most significant invention in the history of coatings, due to the immeasurable benefits in odor, toxicity, flammability, ease of handling, cleanup, and often performance, of emulsions, compared to solvent- or oil-based products. In the new water-based coatings, it is the microscopic emulsion particles which coalesce, act as the binder that holds the pigment particles together, form a continuous film, and are the glue, sticking to the substrate. Emulsion particles are essentially tiny plastic particles which are dispersed in water.

Coatings emulsions are generally formed by addition polymerization of common, highly available monomers, using free radical initiators to create polymers having molecular weights from a few thousand up to millions. The polymerization is most often stabilized by non-ionic and/or anionic surfactants, which emulsify the insoluble monomer droplets, and then stabilize the resulting particles, usually in the shape of a sphere. In addition to surfactants, emulsions are sometimes stabilized with water-soluble poly-

Continued

mers, which act as resin support for the growing polymer particles. Also, many coatings emulsion polymers contain ionic groups which enhance stability via contributing to an electrical double-layer.

Commercially significant coatings emulsions include: acrylics: copolymers of acrylates, such as butyl acrylate, and methacrylates, such as methyl methacrylate; styrene-acrylics: copolymers of styrene with an acrylate monomer; and vinyl acetate polymers: homopolymers of vinyl acetate, or copolymers with softer monomer such as ethylene or butyl acrylate.

Polymers which are 100% acrylic are known for their outstanding exterior durability properties, as well as excellent alkali resistance and overall high performance. Styrene is generally lower-cost than many other monomers, so styrene-acrylics are lower in cost than all acrylics but have poorer exterior durability because styrene is a UV absorber and degrades. Styrene-acrylics, do, however, generally have very good water-resistance properties due to their hydrophobicity. Styrene acrylics are popular in some areas of Europe, Asia, and Latin America. Vinyl acetate is also somewhat lower in cost, so these polymers are popular for interior paints, particularly in the United States, where exterior durability and alkali resistance are not performance issues.

Most emulsion polymers are spheres, generally the lowest-energy and therefore most stable configuration. However, there are other particle shapes and morphologies which can be obtained during emulsion poly-

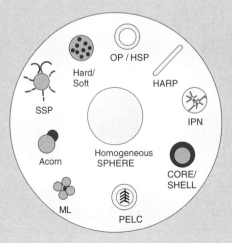

OP/HSP – Opaque Polymer
HARP – High Aspect Ratio Polymer
IPN – Interpenetrating Network
PELC – Polymer Encapsulated Latex
ML – Multilobe™
SSP – Soluble Shell Polymer

Figure 6.14

merization, when special properties are desired which can be achieved via a unique morphology. Several alternative morphologies are shown in Figure 6.14.

Dr John M. Friel
Rohm & Haas
Research Laboratories
Philadelphia
USA

Sample problems

1. Calculate the Debye lengths for 0.1 mM, 10 mM and 100 mM aqueous solutions of NaCl and $MgSO_4$, assuming that the salts are completely ionized.

2. Calculate and graph the concentration profiles for Na^+ and Cl^- ions next to a planar charged surface with a potential of -85 mV immersed in a 10 mM NaCl solution.

3. Show graphically how the surface charge density varies with the surface potential for a planar surface in different Debye-length solutions.

4. Use the superposition principle to calculate the electrostatic swelling pressure generated between parallel clay platelets with surface potentials of -110 mV, at a separation of 35 nm in a 1.5 mM aqueous solution of NaCl at 25 °C.

5. Estimate the thickness of a water film of 0.1 mM NaCl solution on a glass plate, 1 cm above a water reservoir. Assume that the water completely wets the glass and that the water/glass interface has an electrostatic potential of -60 mV and that only gravitational and electrical double-layer forces need be considered. Also assume that there is no surface charge at the water/air interface.

6. The surface of a colloid dispersed in aqueous salt solution was found to have an equilibrium surface electrostatic potential of +80 mV, due to the specific adsorption of Na^+ ions. What is the (chemical) free energy of adsorption of Na^+ ions to this surface?

7. Simplify (6.13) to give a simple physical interpretation of the Debye length.

8. Estimate the concentration of Na^+ ions at the centre layer of an aqueous soap film drawn from a mM NaCl solution, if the electrostatic potential at each surface is −20 mV and the film is 10 nm thick.

9. The electrostatic potential at a distance of 5 nm away from a charged, flat surface was found to be −10 mV in an aqueous 0.1 mM NaCl solution. Estimate the electrostatic potential at the surface. What are the concentrations of each ion at this distance away from the surface? Estimate the osmotic pressure at this plane.

Experiment 6.1 Zeta potential measurements at the silica/water interface

Introduction

The stability of most colloidal solutions depends critically on the magnitude of the electrostatic potential (ψ_0) at the surface of the colloidal particle. One of the most important tasks in colloid science is therefore to obtain an estimate of ψ_0 under a wide range of electrolyte conditions. In practice, the most convenient method of obtaining ψ_0 uses the fact that a charged particle will move at some constant, limiting velocity under the influence of an applied electric field. Even quite small particles (i.e. <1 μm) can be observed using a dark-field illumination microscope and their (average) velocity directly measured. The technique is called 'microelectrophoresis'.

At low electric fields [O(1 V/cm)] the speed (U) of the particles is directly proportional to the applied field (E) and hence we can define a parameter called the electromobility (μ) of the particles, given by U/E. Using the Poisson–Boltzmann theory of the diffuse electrical layer next to a charged surface, a simple relationship between μ and ψ_0 can be derived. However, because of doubts about the validity of the theory we introduce another surface potential called the 'zeta potential' (ζ) to represent the surface potential obtained from the electromobility measurements. This potential corresponds to the electrostatic potential at the plane of shear in the liquid, which is assumed to be close to the particle's surface. (Note that in general it usually reasonable to assume that $\psi_o = \zeta$.)

This potential is given by the Smoluchowski equation

$$\zeta = \frac{\mu\eta}{\varepsilon_0 D} \qquad (6.53)$$

where η is the viscosity of the solution, ε_0 the permittivity of free space and D the dielectric constant. It is important to note that because of the method of derivation of this equation it is only valid for colloidal particles which are large compared with the Debye length (κ^{-1}) of the electrolyte solution (i.e. $\kappa a \gg 1$, where a is the radius of the colloidal particles). In general, therefore, this equation will be valid for particles of $1\,\mu m$ or larger radius.

The surface of a silica (or glass) particle contains a high density of silanol groups (about 1 per $25\,\text{Å}^2$) which dissociate to some extent in water to give a negatively charged surface, Figure 6.15.

The magnitude of the electrostatic potential at the silica surface is, as expected from the law of mass action, pH-dependent. The variation in surface (or zeta) potential with pH must therefore be dependent on the dissociation constant of the surface silanol (Si-OH) group.

In this experiment a 'zeta-meter' is used to determine the variation in the zeta potential of silica at constant pH (5.7) over a range of concentrations of a cationic surfactant CTAB, which should adsorb on the ionized silanol groups on the silica particle surface.

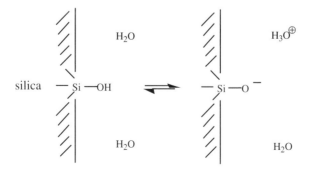

Figure 6.15   Ionization of the surface of silica in water.

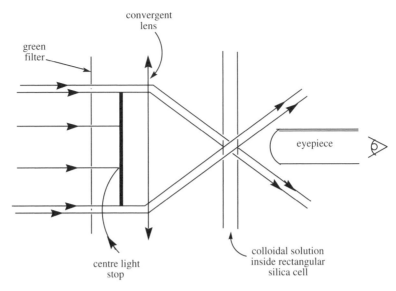

Figure 6.16 Diagram of the dark-field illumination system used to visualise colloidal particles in water.

Experimental details

This experiment is based on the use of a classic Rank Bros (Cambridge, UK) Mark 2 microelectrophoresis apparatus, which is based on the manual microscopic observation of the speed of colloids, detected by dark-field illumination, under the action of an applied electric field. Either a rectangular or a cylindrical quartz cell can be use. For a comprehensive reference text, see R.J. Hunter, *The Zeta Potential* (1981). Before using the zeta-meter apparatus make up colloidal solutions of silica using about $0.01\,g$ solid per $100\,mL$ of a range of CTAB solutions from $10^{-6}\,M$ to $10^{-2}\,M$, in $10^{-2}\,M$ KBr (to keep the Debye length constant). Always shake these solutions thoroughly before transferring to the apparatus cell.

An illustration of the illumination used in this type of apparatus, to observe the motion of colloid particles, is given in Figure 6.16. Basically, the illumination system and microscope allow you to observe the motion of the silica particles, which are seen as bright star-like objects on a green background. When an electric field is applied the average time taken for the particles to travel a distance of one square on the eyepiece graticule can be easily measured. One particle is measured each time the field is applied for a short time (i.e. 10–20 s) and the polarity is then reversed and the speed in the opposite direction measured. The

Quartz cell

Lamp source

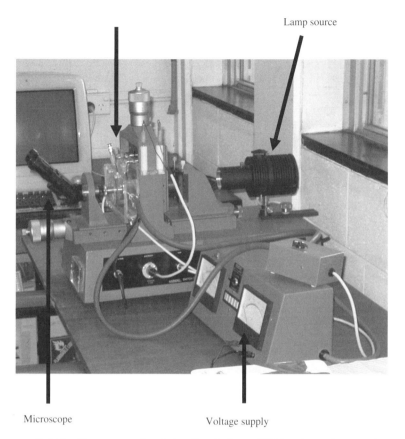

Microscope

Voltage supply

Figure 6.17   Photograph of the Rank Bros MK2 microelectrophoresis instrument.

polarity must be reversed **each time** and the field never left on for longer than about 30 s, so that the possibility of polarization effects is reduced. Usually, between 10 and 20 particles in the field of view are measured and the average value obtained. The applied voltage should be varied to make the particles move over one square in about 10 s but this voltage must never be increased above about 30 V. Be careful only to measure particles clearly in the plane of focus, since the microscope will have been set to measure at the fluid stationary plane within the rectangular cell. Particles not in focus will travel at a speed which will include a fluid flow component and will lead to experimental errors if included in the data.

A photograph of a typical Rank Bros instrument is given Figure 6.17. Once the apparatus is set up, all you have to do is change the colloidal

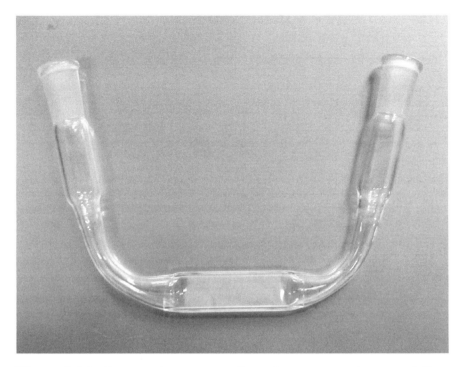

Figure 6.18 Rectangular quartz cell used to measure electromobility.

solutions in the rectangular cell and measure their mobilities. Be careful to check the direction the particles move in, which denotes their sign of charge. This sign will change at some CTAB concentration. (As a guideline, in water the silica particles are negatively charged.)

In order to change solutions, the electrodes are first carefully removed (very little effort should be required to do this – be careful, these cells are expensive) and are then stored in distilled water. The old solution is aspirated out and the cell rinsed and aspirated with distilled water. Finally, the new solution is poured into the cell and the electrodes carefully replaced so that no air bubbles are trapped near the electrodes. The new solution should be left to equilibrate (at 25 °C) for about 15 minutes before measurement. A typical rectangular quartz cell is shown in Figure 6.18. The speed of the colloid particles is measured at a stationary fluid layer within the rectangular cell. These cells are expensive and must be handled carefully. At the end of the experiment the electrodes should be removed and placed in distilled water and the cell should be rinsed and filled with distilled water.

The speed of the particles can be easily calculated using the fact that each square of the graticule corresponds to a known distance, typically

$60\,\mu m$. The electric field applied is also simply calculated by dividing the applied voltage directly measured on the instrument by the distance between the electrodes. The effective inter-electrode distance is obtained by measuring the conductance of a standard electrolyte solution, say 0.01 M KCl, in the cell. The distance is calculated from the known conductivity of the solution. The average electromobility of the colloids is thus obtained and hence, using (6.53), the average zeta-potential.

Plot your results on a graph of zeta potential in mV as a function of CTAB concentration.

Questions

1. Your basic data are in the form of time (in seconds). In determining the average speed of the particles should you first average the time intervals and then invert to calculate the average speed or should you invert the times (to give the speed) and then find the average speed?

2. At what concentration of CTAB was the silica surface uncharged?

3. Draw a schematic diagram of the type of CTAB adsorption you would expect at each concentration.

4. Would you expect CTAB adsorption to increase further with concentration above $10^{-3}\,M$?

7

Van der Waals Forces and Colloid Stability

Historical development of van der Waals forces. The Lennard-Jones potential. Intermolecular forces. Van der Waals forces between surfaces and colloids. The Hamaker constant. The DLVO theory of colloidal stability.

Historical development of van der Waals forces and the Lennard-Jones potential

In 1873 van der Waals pointed out that real gases do not obey the ideal gas equation $PV = RT$ and suggested that two 'correction' terms should be included to give a more accurate representation, of the form: $(P + a/v^2)(V - b) = RT$. The term a/v^2 corrects for the fact that there will be an attractive force between all gas molecules (both polar and non-polar) and hence the observed pressure must be increased to that of an ideal, non-interacting gas. The second term (b) corrects for the fact that the molecules are finite in size and act like hard spheres on collision; the actual free volume must then be less than the total measured volume of the gas. These correction terms are clearly to do with the interaction energy between molecules in the gas phase.

Applied Colloid and Surface Chemistry Richard M. Pashley and Marilyn E. Karaman
© 2004 John Wiley & Sons, Ltd.   ISBN 0 470 86882 1 (HB)   0 470 86883 X (PB)

In 1903 Mie proposed a general equation to account for the interaction energy (V) between molecules:

$$V = -\frac{C}{d^m} + \frac{B}{d^n} \tag{7.1}$$

of which the most usual and mathematically convenient form is the Lennard-Jones 6-12 potential:

$$V_{LJ} = -\frac{C}{d^6} + \frac{B}{d^{12}} \tag{7.2}$$

where the first term represents the attraction and the second the repulsion between two molecules separated by distance d. This equation quite successfully describes the interaction between non-polar molecules, where the attraction is due to so-called dispersion forces and the very short-range second term is the Born repulsion, caused by the overlap of molecular orbitals.

From our observation of real gases it is clear that attractive dispersion forces exist between all neutral, non-polar molecules. These forces are also referred to as London forces after the explanation given by Fritz London in about 1930. At any given instant a neutral molecule will have a dipole moment because of fluctuations in the electron distribution in the molecule. This dipole will create an electric field which will polarize a nearby neutral molecule, inducing a correlated dipole moment. The interaction between these dipoles leads to an attractive energy of the form $V = -C/d^6$. The time-averaged dipole moment of each molecule is, of course, zero but the time-averaged interaction energy is finite, because of this correlation between interacting, temporary, dipoles. It is mainly this force which holds molecular solids and liquids, such as hydrocarbons and liquefied gases, together. The L-J interaction potential V between molecules of a liquid (or solid) separated by distance d, is illustrated in Figure 7.1. In this case the molecules in the liquid would have an equilibrium spacing of d_m. Further, if we knew the detailed 'structure' of the liquid, that is, the radial distribution function $g(r)$, we could calculate its internal energy $<U>$ and hence relate molecular interactions directly to the systems thermodynamic parameters. This calculation can be performed using the classical Hamiltonian function of the sum of the internal (interaction) energy and the molecular kinetic energy, $3NkT/2$,

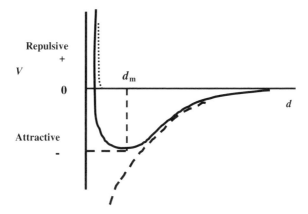

Figure 7.1 Interaction energy between two molecules.

$$\langle U \rangle = \frac{3}{2}NkT + \frac{1}{2}N\rho \int_0^\infty 4\pi r^2 g(r)V(r)dr \qquad (7.3)$$

and if we know how $g(r)$ varies with temperature, then we can calculate $<U>$ as a function of temperature, from which other thermodynamic parameters can be generated.

It is interesting to note that molecular interaction forces vary with the inverse distance to a power greater than 3. This means that these forces are short-ranged within a material, unlike gravitational forces, and only molecules within 10 or so diameters effectively contribute to the cohesive or surface energy of a material. This is quite unlike gravity, which has a slower distance dependence and so has to be summed over the entire body or region of space. Interestingly, this difference was not appreciated by Einstein in some of his early work.

Because the dispersion force acts between neutral molecules it is ubiquitous (compare the gravitational force); however, between polar molecules there are also other forces. Thus, there may be permanent dipole–dipole and dipole-induced dipole interactions and, of course, between ionic species there is the Coulomb interaction. The total force between polar and non-polar (but not ionic) molecules is called the *van der Waals force*. Each component can be described by an equation of the form $V = C/d^n$, where for the dipole–dipole case $n = 6$ and C is a function of the dipole moments. Clearly, it is easy to give a reasonable distance dependence to an interaction; however, the real difficulty arises in determining the value of C.

Common types of interactions between atoms, ions and molecules in vacuum are given in Table 7.1. In the table, $V(r)$ is the interaction free

Table 7.1

Type of interaction	Interaction energy $V(r)$
Covalent	Complicated, short-range
Charge–charge	$\dfrac{q_1 q_2}{4\pi\varepsilon_0 r}$ Coulomb's law
Charge (q)–dipole (u), freely rotating dipole	$\dfrac{-q^2 u^2}{6(4\pi\varepsilon_0)^2 kTr^4}$
Dipole–dipole, both freely rotating	$\dfrac{-u_1^2 u_2^2}{3(4\pi\varepsilon_0)^2 kTr^6}$ (Keesom energy)
Charge (q)–non-polar (α)	$\dfrac{-q^2 \alpha}{2(4\pi\varepsilon_0)^2 r^4}$
Rotating dipole (u)–non-polar(α)	$\dfrac{-u^2 \alpha}{(4\pi\varepsilon_0)^2 r^6}$ (Debye energy)
Two non-polar molecules	$\dfrac{-3h\nu\alpha^2}{4(4\pi\varepsilon_0)^2 r^6}$ (London dispersion energy)
Hydrogen bond	Complicated, short-range, energy roughly proportional to $-1/r^2$

energy (in J); q is the electric charge (C); u, the electric dipole moment (Cm); α, the electric polarizibility ($C^2m^2J^{-1}$) and r is the distance between interacting atoms or molecules (m). ν is the electronic absorption (ionization) frequency (s^{-1}). The corresponding interaction force is, in each case, obtained by differentiating the energy $V(r)$ with respect to distance r.

A more sophisticated model for water molecule interactions is given by the four-point charge model. The interaction potential has the form

$$V(X_1, X_2) = V_{LJ}(R_{12}) + S(R_{12})V_{HB}(X_1, X_2) \qquad (7.4)$$

where X is the generalized coordinate of molecule 1 specifying the position and orientation of that particle. $V_{LJ}(R_{12})$ is a Lennard-Jones 6:12 potential with parameters for neon, which is iso-electronic with water, and V_{HB} is a slight variant of the Bjerrum four-point charge model (Figure 7.2). The function $S(R_{12})$ is a switching function to give small weight to configurations in which the two point charges on neighbouring molecules overlap. The modified four-point charge model of

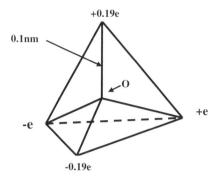

Figure 7.2 Bjerrum four point charge model for water.

Bjerrum has been used for molecular dynamics simulations of liquid water, to calculate its thermodynamic and structural properties. It turns out that this very important liquid is extremely difficult to model theoretically.

Dispersion forces

It is instructive to follow the derivation of the London dispersion interaction, for the simplest case of two interacting hydrogen atoms, using the Bohr model where the electron is regarded as travelling in well-defined orbits about the nucleus. The orbit of smallest radius, a_0, is the ground state and Bohr calculated that

$$a_0 = \frac{q^2}{8\pi\varepsilon_0 h\nu} \tag{7.5}$$

where ν is the characteristic frequency (in Hz) associated with the electron's motion around the nucleus. (Note that the value of a_0 corresponds to the maximum value in the electron density distribution $|\psi|^2$ in the electronic ground state of hydrogen as calculated from quantum mechanics.) The energy $h\nu$ is the first ionization potential of the H atom. Consider the H atom as having a spontaneously formed dipole moment

$$P_1 = a_0 q \tag{7.6}$$

The electric field E due to this instantaneous dipole at distance R will be

$$E \cong \frac{P_1}{4\pi\varepsilon_0 R^3} \cong \frac{a_0 q}{4\pi\varepsilon_0 R^3} \tag{7.7}$$

(which can be calculated simply from Coulomb's law). If a neutral molecule is at R, it will be polarized, producing a dipole moment depending on its polarizability α; thus

$$P_2 = \alpha E \cong \frac{\alpha a_0 q}{4\pi\varepsilon_0 R^3} \tag{7.8}$$

α measures the ease with which the electron distribution can be displaced and is proportional to the volume of the atom (H in this case):

$$\alpha \cong 4\pi\varepsilon_0 a_0^3 \tag{7.9}$$

The potential energy of interaction of the dipoles P_1 and P_2 is then

$$V(R) = -\frac{P_1 P_2}{4\pi\varepsilon_0 R^3} \quad \text{(for fixed and aligned diploes)}$$

$$= -\frac{\alpha a_0^2 q^2}{(4\pi\varepsilon_0)^2 R^6}$$

$$\therefore V(R) = -\frac{2h\nu\alpha^2}{(4\pi\varepsilon_0)^2 R^6} \tag{7.10}$$

Thus, the dispersion force interaction depends very much on the polarizability or response of the molecule to an electric field.

Retarded forces

Before we move on to consider the interaction between macroscopic bodies, let us look briefly at the phenomenon of 'retardation'. The electric field emitted by an instantaneously polarized neutral molecule takes a finite time to travel to another, neighbouring molecule. If the molecules are not too far apart the field produced by the induced dipole will reach the first molecule before it has time to disappear, or perhaps form a dipole in the opposite direction. The latter effect does, however, occur at larger separations (>5 nm) and effectively strengthens the rate of decay with distance, producing a dependence of $1/R^7$ instead of $1/R^6$.

At closer separations, where the van der Waals forces are strong, the interaction is non-retarded and we will assume this is the case from here onwards.

Van de Waals forces between macroscopic bodies

In colloid and surface science we are interested in calculating the van der Waals interaction between macroscopic bodies, such as spherical particles and planar surfaces. If the dispersion interaction, for example, were additive we could sum the total interaction between every molecule in a body with that in another. Thus, if the separation distance between any two molecules 'i' and 'j' in a system is

$$R_{ij} = \left| \vec{R_j} - \vec{R_i} \right| \tag{7.11}$$

then a sum of the interaction energy between all molecules gives:

$$V^{1,2\ldots N} = 0.5 \sum_{i=1}^{N} \sum_{j=1(\neq i)}^{N} V^{i,j}(R_{ij}) \tag{7.12}$$

where $V^{ij}(R_{ij})$ is the interaction energy between molecules i and j in the *absence* of all other molecules. However, this approach is only an approximation because the interaction is dependent on the presence of other molecules and the correct treatment has to include many-body effects. A complete but rather complicated theory which includes these effects is called the Lifshitz theory, and was derived via quantum-field-theoretic techniques. Although the complete equation is complicated we can represent it in a simple and approximate form when the interaction is non-retarded (at separations less than about 5 nm). Thus, for the case of two planar macroscopic bodies we have Figure 7.3. The van der Waals interaction energy per unit area is given by

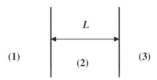

Figure 7.3 Diagram of two planar surfaces separated by distance L.

$$V_{123}^F(L) = -\frac{A_{123}}{12\pi L^2} \tag{7.13}$$

where A_{123} is the so-called 'Hamaker constant', which is positive when the interaction is attractive. Similarly, between identical spherical particles, we have Figure 7.4.

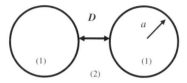

Figure 7.4 Diagram of two colloidal spheres separated by distance D.

$$V_{121}^S(D) = -\frac{A_{121}a}{12D} \tag{7.14}$$

These equations are simple and become powerful if we know the value of A for particular materials. Tables of A values which have been calculated using the Lifshitz theory are available and some examples are given in Table 7.2.

Table 7.2

Components	Hamaker constant
	A $(10^{-20}\,\text{J})$
water/vacuum/water	3.7
polystyrene/vacuum/polystyrene	6.1–7.9
silver/vacuum/silver	40
quartz/vacuum/quartz	8–10
hydrocarbon/vacuum/hydrocarbon	6
polystyrene/water/polystyrene	1.3
quartz/water/quartz	1.0
dodecane/water/dodecane	0.5
Teflon/water/Teflon	0.3

Theory of the Hamaker constant

Examination of Table 7.2 reveals some interesting features, such as the effect of the medium in between two macroscopic bodies, which clearly has a marked effect on reducing the van der Waals attraction. This effect can be understood if we examine the difference between vacuum and a dielectric as the medium in which we bring two bodies together. In the first case, the original body has no molecules to interact with outside itself, and hence lower its free energy, whereas in the second the body already interacts with molecules in the surrounding medium which are merely replaced by those in the approaching second body. The magnitude of the Hamaker constant in the second case will clearly depend on the interaction between all the components. This is also the reason why any two dielectrics will be attracted by van der Waals forces in a vacuum but not necessarily when immersed in some other medium.

Although it is reasonable to simply use the values calculated by theoreticians (and in a few cases measured by experimentalists) for the Hamaker constant, it is important to understand something about how it is calculated. The Hamaker constant is, in fact, a complicated function of the frequency-dependent dielectric properties of all the media involved. The way in which the varying electric fields generated by one body interact with another determine the van der Waals interaction. In order to understand this let us look at the effect of placing a dielectric material (i.e. $\varepsilon > \varepsilon_0$) between two charged metal (conducting) plates initially in a vacuum. This situation is illustrated in Figure 7.5.

One of the fundamental laws of electrostatics is that due to Gauss:

$$\int \vec{E} \cdot \hat{n} \mathrm{d}a = \frac{Q}{\varepsilon_0} \tag{7.15}$$

which may be derived from Maxwell's equations (see earlier) and says that the total electric flux' through (normal to) a surface is directly related to the total charge Q inside the surface. Let us use this general

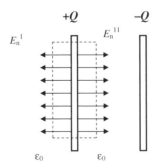

Figure 7.5 Electric field around two charged plates of a capacitor.

equation to study one half of the capacitor system illustrated, where the total area is shown as a dashed line (Figure 7.5).

The electric flux normal to the surface is given by:

$$(E'_n + E''_n) = \frac{Q}{A\varepsilon_0} \tag{7.16}$$

where A is the area of one side of the capacitor plate and Q is the total charge on the plate. However, since the electric fields are symmetrical around a flat plate $E'_n = E''_n$ and hence $E = Q/2A\varepsilon_0$. This, if we now also include the field due to the other negative plate, becomes

$$E = Q/A\varepsilon_0 \tag{7.17}$$

Since the capacitance, C, is defined as

$$C = Q/V \tag{7.18}$$

where V is the potential difference between the plates, and this is the work done in moving a unit positive charge from one plate to the other, that is,

$$Ed = V \tag{7.19}$$

(since the field E is the force acting on a unit positive charge inside the capacitor plates) we can obtain the simple result that

$$C_0 = \frac{\varepsilon_0 A}{d} \tag{7.20}$$

where the capacitance depends only on the area and the separation of the plates and the permittivity of the medium in between the plates. What happens if we now fill this free space with a dielectric material? Since for any dielectric $\varepsilon_D > \varepsilon_0$ we can see immediately that the capacitance will increase via the formula

$$\frac{C_D}{C_0} = \frac{\varepsilon_D}{\varepsilon_0} \tag{7.21}$$

This is a fundamental property of dielectrics and means that if the charge on the plates is fixed, the potential difference between the plates and hence the electric field inside (in the dielectric) must have fallen

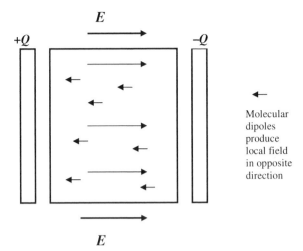

Figure 7.6 Effect of a dielectric material on the electric field within a capacitor.

compared with that in free space. How this happens is explained in the schematic Figure 7.6. In the electric field between the plates, the dielectric material polarizes, for example, by a change in electron distribution in the molecules or, if polar, by re-orientation of the dipoles (especially for a liquid such as water). The induced dipolar field then must act in the opposite direction and reduces the total electric field and the potential difference. Thus the dielectric material between the plates of the capacitor increases its capacitance. The dielectric constant of the material can be measured in this way.

Now let us examine what would happen to the response of the dielectric if we put an alternating voltage on the capacitor of frequency ω. If ω is low (a few Hz) we would expect the material to respond in a similar manner to the fixed-voltage case, that is ε_D (static) $= \varepsilon(\omega) = \varepsilon(0)$. (It should be noted that ε_0, the permittivity of free space, is not frequency-dependent and that $\varepsilon(0)/\varepsilon_0 = D$, the static dielectric constant of the medium.) However, if we were to increase ω to above microwave frequencies, the rotational dipole response of the medium would disappear and hence $\varepsilon(\omega)$ must fall. Similarly, as we increase ω to above IR frequencies, the vibrational response to the field will be lost and $\varepsilon(\omega)$ will again fall. Once we are above far-UV frequencies, all dielectrics behave much like a plasma and eventually, at very high values, $\varepsilon(\omega)|_{\omega \approx \infty} = 1$.

What is actually happening at the frequencies ω_i, where there is a sudden reduction in the response of the dielectric? We can, in fact, treat the interaction of varying electric fields with dielectrics as though the latter were made up of electron or dipole oscillators, such that when the resonant frequency is reached, electric energy is absorbed and, usually, dissipated as heat. The ω_i values must then correspond to the absorption peaks of the material. In order to represent this behaviour we allow the frequency-dependent dielectric constant to have an imaginary component, thus:

$$\varepsilon(\omega) = \varepsilon'(\omega) + i\varepsilon''(\omega) \tag{7.22}$$

where $\varepsilon''(\omega)$ is directly proportional to the absorption coefficient of the dielectric. At frequencies where there is no absorption it follows that

$$\varepsilon(\omega) = \varepsilon'(\omega) = n(\omega)^2 \tag{7.23}$$

where $n(\omega)$ is the refractive index (at that frequency). From what we have said about dielectrics, we would expect something like Figure 7.7 for the behaviour of $\varepsilon'(\omega)$ and $\varepsilon''(\omega)$, with frequency of the electric field. If the absorption data, that is $\varepsilon''(\omega)$ over a very wide frequency range – in practice from microwave to far-UV, has been measured we can use the Lifshitz equation to calculate the Hamaker constant. Unfortunately, these data are known in detail for only a few materials (such as water and polystyrene) and so oscillator models for the main absorption peaks (say in the IR and UV) are often constructed and used to calculate $\varepsilon''(\omega)$.

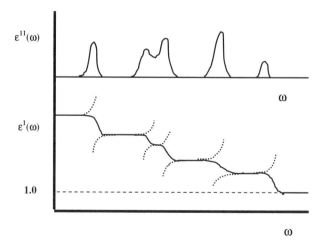

Figure 7.7 Typical responses for the real and imaginary components of the dielectric constant to frequency.

We can, however, make some semi-quantitative comments about the type of van der Waals forces we expect from the main absorption peaks and the refractive index of transparent dielectrics. For example, if two dielectric bodies which interact through vacuum have very similar absorption spectra, the van der Waals attraction will be strong. Also, if the intervening medium has a spectrum similar to that of the interacting bodies the attraction will be weak (and can even be repulsive).

Even if we know nothing about the absorption properties we can still deduce something about the magnitude of van der Waals forces from the refractive index at visible frequencies, where it is known for transparent (i.e. non-absorbing) materials. This is because a high refractive index must mean that there is still a substantial dielectric response at these (visible) frequencies and therefore there must be further absorption at higher frequencies in the UV range (since $\varepsilon(\omega) = n(\omega)^2$ where the material is transparent). Absorption in the UV range is very important for short-range van der Waals interactions ($d < 5$ nm), where these frequencies are not retarded, hence the refractive index is related to the Hamaker constant. This is illustrated by Table 7.3, which lists refractive indices of materials and the corresponding Hamaker constants for interactions across a vacuum.

As an aside, it should be noted that even when electromagnetic radiation (e.g. light) passes through a transparent dielectric medium, the varying electric fields polarize the molecules and these re-emit radiation of the same wavelength but with a phase shift. It is the combination of this re-emitted field with the original which gives a transmitted beam the appearance of travelling at a slower speed, C_D, compared with its speed in a vacuum, C. The refractive index can be shown from this approach to be numerically equal to the ratio C/C_D. The electric field 'in between' the molecules of the dielectric still travels with the speed

Table 7.3

Material	n_D ($\lambda = 589.3$ nm)	A_{11} (10^{-20} J)
water	1.33	3.7
hexadecane	1.43	5.2
CaF_2	1.43	7.2
fused silica (SiO_2)	1.46	6.6
crystalline quartz (SiO_2)	1.54	8.8
mica (aluminosilicate)	1.55	10
calcite ($CaCO_3$)	1.5–1.7	10.1
sapphire (Al_2O_3)	1.77	16

of light in a vacuum. Absorption occurs when the interacting field is not re-emitted and this occurs at the resonant frequencies of the material, where the dissipated electric field energy usually appears as heat.

Use of Hamaker constants

Once we have established reasonable values for the Hamaker constants we should be able to calculate, for example, adhesion and surface energies, as well as the interaction between macroscopic bodies and colloidal particles. Clearly, this is possible if the only forces involved are van der Waals forces. That this is the case for non-polar liquids such as hydrocarbons can be illustrated by calculating the surface energy of these liquids, which can be directly measured. When we separate a liquid in air we must do work W_C (per unit area) to create new surface, thus:

$$V_{11} = -W_c = -\frac{A_{11}}{12\pi L^2} \tag{7.24}$$

If we assume a reasonable separation distance in the liquid, of say $L = 0.15$ nm, then we calculate a liquid surface tension value, γ, of $30\,\text{mJ}\,\text{m}^{-2}$, whereas the measured value is $28\,\text{mJ}\,\text{m}^{-2}$, which is reasonably close. However, if we carry out the same calculation for water using the same spacing value then $\gamma = 22\,\text{mJ}\,\text{m}^{-2}$, which is much lower than the measured value of about $72\,\text{mJ}\,\text{m}^{-2}$. The main reason for this large discrepancy is that the surface energy of water is high because of the short-range, structural hydrogen-bonding between water molecules. At the water/air interface these molecules are orientated quite differently from those in bulk. Of course, we have included dipole–dipole interactions in our calculation of the Hamaker constant (by using the microwave and zero-frequency contributions) but these are the bulk properties of liquid water, which do not represent structural and orientational changes in the water dipoles at the surface of the liquid. Application of the theory to calculate surface tension values clearly works best for simple liquids.

The DLVO theory of colloid stability

Although for some interfaces it can be difficult to calculate the surface energy in this way, the Hamaker constant does afford us a powerful

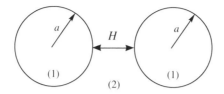

Figure 7.8 Two interacting identical colloids.

tool for the calculation of the attractive forces acting between colloids. Typically, these forces are opposed by the charge on the particle surface and the generation of a repulsive stabilizing force, called the 'electrical double-layer repulsion', discussed in Chapter 6. Combination of the simple equation obtained in the present chapter (7.14) for the van der Waals attraction between spherical colloids, with that derived in Chapter 6 for the double-layer repulsion (6.52), leads to a simplified form of the DLVO theory for the stability of colloidal solutions. Thus, for the general case of two interacting spherical colloids illustrated in Figure 7.8, the total interaction energy can be given approximately by the relation

$$V_{\mathrm{T}} \approx \left[2\pi a \varepsilon_0 D \psi_0^2 \exp(-\kappa H)\right] - \frac{a A_{121}}{12H} \tag{7.25}$$

Some typical interaction curves are given in Figure 7.9.

One of the first accurate experimental tests of the DLVO theory was published by one of the present authors (Pashley, 1981). To obtain the required accuracy, the forces were measured not between colloids but between two molecularly smooth crystals of muscovite mica. The results obtained were in almost perfect agreement with theory (Figure 7.10). The measured experimental points (large black dots in the figure, actually values of force F scaled by the radius of the surfaces R) were found to agree with the DLVO predictions (dashed line) at all separations. The long range of the forces was due to the low electrolyte concentration in distilled water, giving an effective Debye length of 145 nm. At small separations the inset shows that the surfaces were pulled into adhesive contact by van der Waals forces, as predicted by the theory.

In 1991, the same author helped to develop a new experimental procedure, called the 'colloid probe technique', which is now widely used to measure the interaction forces between colloidal surfaces (see Ducker et al., 1991).

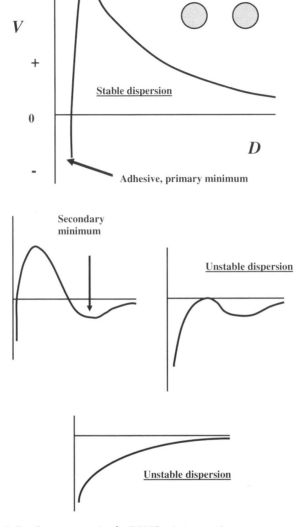

Figure 7.9 Some typical DLVO interaction curves for colloidal solutions under different conditions.

Flocculation

For colloidal solutions, as a general rule, a barrier of 15–25 kT is sufficient to give colloid stability, where the Debye length is also relatively large, say, greater than 20 nm. This electrostatic barrier is sufficient to

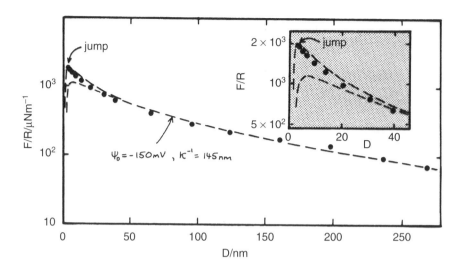

Figure 7.10 Measured DLVO fones between two smooth mica surfaces in water.

maintain meta-stability of the colloidal dispersion but the system is usually not thermodynamically stable and even a low solubility of the dispersed phase will allow growth of large particles at the expense of small ones, as expected from the interfacial energy at the particle surface. This growth is called 'Ostwald ripening'. The rate at which it occurs will depend on the solubility of the disperse phase in the dispersion medium. Note that even silica has a finite solubility in water.

If the electrostatic barrier is removed either by specific ion adsorption or by addition of electrolyte, the rate of coagulation (often followed by measuring changes in turbidity) can be described fairly well from simple diffusion-controlled kinetics and the assumption that all collisions lead to adhesion and particle growth. Overbeek (1952) has derived a simple equation to relate the rate of coagulation to the magnitude of the repulsive barrier. The equation is written in terms of the stability ratio:

$$W = \frac{R_f}{R_s}$$

where R_f is the rate of fast coagulation, that is, where there is no barrier and all collisions are successful, and R_s is the rate of slow coagulation against a barrier. The dependence of W on the barrier is given by the relation

$$W = \frac{1}{2\kappa a} \exp\left(\frac{V_{max}}{kT}\right)$$

Thus, for a barrier of $20\,kT$ and a κa value of 1, the stability ratio, W, has a value of 2.4×10^8, which corresponds to a slow rate of coagulation.

Stability from coagulation is an important property in many industrial processes and products. Often the electrostatic component is not sufficient, for example in high-electrolyte solutions, and it is necessary to generate an additional repulsive barrier between the particles. One commonly used method is to adsorb a water-soluble polymer as a coating around the particles to prevent their close approach. The forces generated are complex and polymer-specific but are classified as steric forces, to distinguish them clearly from the DLVO forces. This is the main reason many biological cells have coatings of natural biopolymers, to give stability in the high level of aqueous electrolyte in animals, of about $0.15\,M\,NaCl$.

An important industrial example of one of the common types of interaction obtained using this equation is for the case of latex particles stabilized by repulsive electrostatic forces. The case shown in Figure 7.11 reflects the situation where rapid coagulation will be generated by even a small increase in electrolyte level by evaporation of water from the paint. Whether stabilized by repulsive electrostatic forces or steric forces, the stable latex paint dispersion must rapidly become unstable when applied as a thin film coating. Typically, the initial latex dispersion contain 50% polymer particles with a diameter of about 0.1 micron, prepared by the emulsion polymerization process described in Chapter 5. Once coated as a thin film, the evaporation of water forces the particles together into a close-packed array and then into a compressed state and finally into a continuous polymer film as illustrated in Figure 7.12. An example of the surface of a drying latex film is shown in the AFM (atomic force microscope) image, Figure 7.13.

Both secondary and primary minimum coagulation are observed in practice and the rate of coagulation is dependent on the height of the barrier. In general, coagulation into a primary minimum is difficult to reverse, whereas coagulation into a secondary minimum is often easily reversed, for example, by diluting the electrolyte. DLVO theory tells us

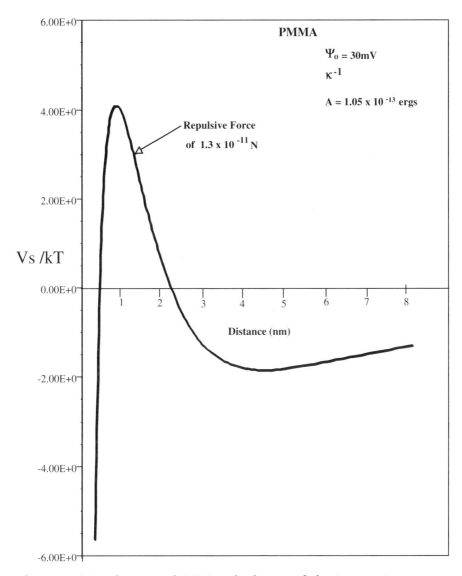

Figure 7.11 Theoretical DLVO calculation of the interaction energy between typical latex particles in water-based paint.

that colloidal solutions will, in general, be stable if the surface potential is high and the electrolyte concentration low, which is well supported by numerous experimental studies. The importance of the concentration and type of electrolyte in determining the stability of colloidal solutions is illustrated in the experimental results in Table 7.4.

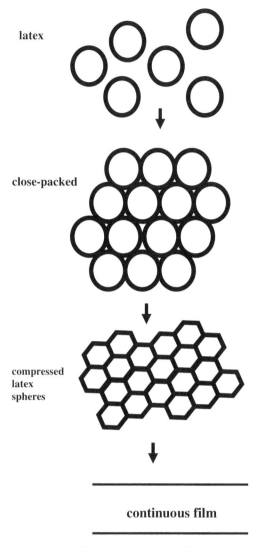

latex

close-packed

compressed
latex
spheres

continuous film

Figure 7.12 Schematic diagram of the film formation process of latex paints.

One of the most effective methods for the coagulation and removal of colloids from solution arises from the use of heterocoagulation, where particles of opposite surface charge are mixed together. A very common and widely used industrial example is afforded by the processing of municipal drinking-water supplies. The most common process uses a coagulant colloid in the form of an initially soluble salt, usually either ferric chloride or alum, which at moderate pH (~8) will

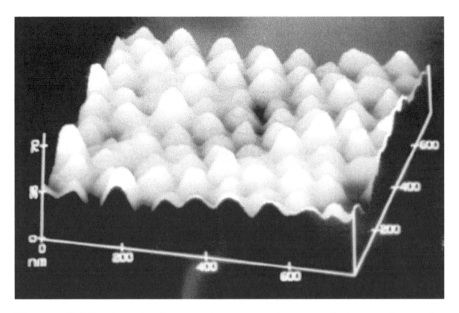

Figure 7.13 Atomic force microscope image of the surface of a drying latex film.

Table 7.4

As$_2$S$_3$ (negatively charged)		Fe$_2$O$_3$ (positively charged)	
Electrolyte	Coagulation concentration (mM)	Electrolyte	Coagulation concentration (mM)
NaCl	51	NaCl	9.25
KCl	49.5	KCl	9.0
KNO$_3$	50	(1/2)BaCl$_2$	9.6
(1/2)K$_2$SO$_4$	65.5	KNO$_3$	12
LiCl	58		
MgCl$_2$	0.72	K$_2$SO$_4$	0.205
MgSO$_4$	0.81	MgSO$_4$	0.22
CaCl$_2$	0.65	K$_2$Cr$_2$O$_7$	0.195
AlCl$_3$	0.093		
Al(NO$_3$)$_3$	0.095		
Ce(NO$_3$)$_3$	0.080		

hydrolyse to form positively charged flocs. Since almost all natural impurities in reservoir water are negatively charged, such as silica, clay and biological cells, the flocs attach to the contaminants and the growing particles are settled and filtered out using high-flow-rate sand-bed filters.

Some notes on van der Waals forces

The degree of correlation in dielectric response between interacting materials leads to some useful generalizations about macroscopic van der Waals forces, thus:

(1) The vdW force is always attractive between any two materials in a vacuum. This is because there is no interaction between a dielectric material and a vacuum. However, in the Casimir effect, the spatial restriction of vacuum quantum fluctuations when two metal plates are placed in close proximity, creates an attractive pressure on them, in addition to the vdw force.

(2) The vdW force between identical materials is always attractive in any other medium. This is because the correlation will always be a maximum for identical materials – since different absorption spectra must always reduce the extent of correlation for different materials.

(3) The vdw force can be repulsive for two different materials interacting within a third medium. In this case one material must interact more strongly with the medium than with the second body.

Industrial Report

Surface chemistry in water treatment

In some instances the raw water reaching water treatment plants may contain pathogens such as the human infectious protozoon *Cryptosporidium parvum*. The environmental form of *C. parvum* is a spheroidal oocyst of 4–6 microns diameter. The oocyst is resistant to conventional chemical disinfectants that are commonly used in water treatment such as chlorine or chloramines. It is therefore essential that *Cryptosporidium* be removed during the coagulation and filtration processes stage in the water treatment train where chlorine and chloramines are relied on as the only disinfectants.

Water treatment by either direct or contact filtration has become common practice for raw water with low turbidity (<3 NTU) and low colour. Simple metal salts such as alum or ferric chloride are added to plant inlet water. Hydrolysis takes place with the formation of hydroxylated species, which adsorb, reducing or neutralizing the charge on the colloidal particles in the raw water, promoting their collision and the formation of flocs that settle or can be filtered out.

In most water filtration plants in Sydney (Australia), that treat approximately 2×10^9 litres/day of raw water, the chemical regimes typically include (i) ferric chloride (up to 5 mg/L), (ii) low-molecular-weight cationic polyelectrolyte as secondary coagulants, and (iii) high-molecular-weight non-ionic polyacrylamide-based flocculants as filter aids. Optimization of the chemical regime in water filtration plants almost invariably delivers acceptable water quality in terms of turbidity (of about 0.15 NTU), colour and trace metals that meets standard water quality guidelines.

Research is continuing on the main factors that influence oocyst flocculation since variations in chemical dosing, water chemistry or oocyst characteristics could potentially lead to oocyst breakthrough in water treatment plants. In recent years researchers have proposed that the interactions between oocysts and different coagulants may be quite different (Bustamante et al., 2001).

Much of the understanding of the appropriate doses of coagulants to use has been developed from empirical success in optimizing water treatment. It is therefore of primary importance to better understand the interaction of oocysts with common coagulants and coagulant aids normally used in water treatment to allow operators to make better informed choices in dosing during changing raw water conditions and to assist in trouble-shooting should problems arise.

Dr Heriberto Bustamante
Research Scientist
Sydney Water Corporation
Sydney, Australia

Sample problems

1. Use schematic diagrams to describe the influence of electrolyte concentration, type of electrolyte, magnitude of surface electrostatic potential and strength of the Hamaker constant on the interaction energy between two colloidal-sized spherical particles in aqueous solution. What theory did you use to obtain your description? Briefly describe the main features of this theory.

2. Two uncharged dielectric materials ('1' and '2') are dispersed as equal-sized, spherical colloidal particles in a dielectric medium, '3'. (see Figure 7.14) If the refractive indices at visible frequencies follow the series: $n_1 > n_3 > n_2$, determine the relative strengths (and sign, i.e. whether repulsive or attractive) of the three possible interactions. Explain your reasoning.

3. Would you expect the van der Waals interaction between two spherical water droplets in air and two air bubbles in water to be the same, if all the spheres are of identical size? Explain your answer.

4. The total DLVO interaction energy (V_S) between two spherical colloids (each of radius a and separated by distance H) is given by the following approximate equation:

$$V_S = \left[2\pi a \varepsilon_0 D \psi_0^2 \exp(-\kappa H)\right] - \frac{a A_{121}}{12H}$$

Calculate the critical value of the surface potential of the colloid which will just give the rapid coagulation case illustrated in Figure 7.15. Assume that the aqueous solution contains 10 mM monovalent electrolyte at 25 °C. Also assume that the Hamaker constant for this case has a value of 5×10^{-20} J.

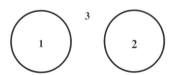

Figure 7.14 Interaction between two different dielectric colloids immered in a third dielectric medium.

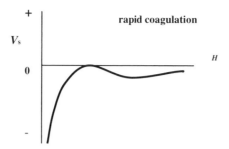

Figure 7.15 Rapid coagulation condition for the interaction between two colloids.

5. Estimate the surface energy of liquid hexane using the fact that the van der Waals interaction for the liquid hexane/vacuum/liquid hexane (flat-surface) case is given by

$$V_{121} = -\frac{A_{121}}{12\pi L^2}$$

where the Hamaker constant A_{121} has a value of 6×10^{-20} J. Assume that the intermolecular spacing within liquid hydrocarbons is about 0.12 nm. How would you expect the Hamaker constant to change if the vacuum were replaced with (a) water and (b) another (immiscible) liquid hydrocarbon?

8

Bubble Coalescence, Foams and Thin Surfactant Films

Thin-liquid-film stability. The effect of surfactants on film and foam stability. Surface elasticity. Froth flotation. The Langmuir trough and monolayer deposition. Laboratory project on the flotation of powdered silica.

Thin-liquid-film stability and the effects of surfactants

Foams are important in many everyday activities and are used in a diverse range of important industrial processes. In food science foams play a major role in both taste and appearance. Personal soaps contain compounds especially designed to stabilize foams, so that the soap can be both retained and easily transferred on to the skin during washing. Too much foam can also be a problem in industrial processes and in home washing machines. The major industrial process of froth flotation is based on the formation of foams for the collection and separation of valuable minerals in large quantities. Flotation is also used to remove earth during the cleaning and processing of vegetables.

When two air bubbles collide in water there is an overall thermodynamic advantage in fusion to form a larger, single bubble. This follows

Applied Colloid and Surface Chemistry Richard M. Pashley and Marilyn E. Karaman
© 2004 John Wiley & Sons, Ltd.   ISBN 0 470 86882 1 (HB)   0 470 86883 X (PB)

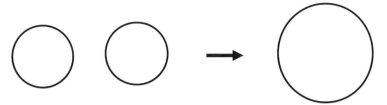

Figure 8.1   Illustration of the reduction in total surface area by the fusion of two air bubbles in water.

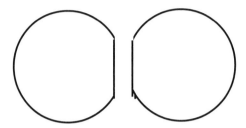

Figure 8.2   Deformation of colliding air bubbles in water.

directly from the reduced interfacial area and hence reduced interfacial energy (Figure 8.1), since at constant gas volume, the radius of the combined bubble must be 1.26 times the radius of the original bubble and it follows that there will be a reduction in interfacial area of just over 20 per cent. This will be the case for all pure liquids and in fact the persistence of bubbles in water is often used as a simple check on its purity. Shaking a flask of water should produce bubbles that collapse within 1–2 s, if the water is clean.

In order to understand the basis for the prevention of bubble coalescence and hence the formation of foams, let us examine the mechanical process involved in the initial stage of bubble coalescence. The relatively low Laplace pressure inside bubbles of reasonable size, say over 1 mm for air bubbles in water, means that the force required to drain the water between the approaching bubbles is sufficient to deform the bubbles as illustrated in Figure 8.2. The process which now occurs in the thin draining film is interesting and has been carefully studied. In water, it appears that the film ruptures, joining the two bubbles, when the film is still relatively thick, at about 100 nm thickness. However, van der Waals forces, which are attractive in this system (i.e. of air/water/air), are effectively insignificant at these film thicknesses.

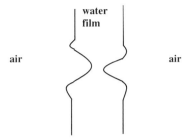

Figure 8.3 Surface correlated wave model for water film rupture.

Most likely, the rupturing occurs because of the correlation and fusion of surface waves at both surfaces of the film, as illustrated in Figure 8.3. It has been suggested that a long-range attractive force, observed between hydrophobic surfaces in water, the so-called hydrophobic interaction, may also operate in thin water films to correlate the peaks in surface waves on the two facing surfaces and so induce fusion. Whatever the mechanism of film rupture, since the coalescence of bubbles is driven by surface energy changes, it is not surprising that adsorption of suitable "surface-active" materials will oppose fusion and create the phenomenon of foaming. Two main factors are necessary to create a stable foam:

1. a repulsive force between the two film surfaces across the water film;

2. a mechanism to induce surface elasticity in the film to protect against mechanical disturbances.

Both of these factors are achieved by the use of surfactants, which will adsorb at the surfaces of the film as illustrated in Figure 8.4. The film (and hence the foam) can be stabilized by an electrostatic repulsive force acting between the ionized, adsorbed surfactants. It can also be stabilized by non-ionic polymeric-type surfactants (e.g. ethylene oxide surfactants), where the large hydrophilic head groups may have the effect of raising the viscosity of the film and hence reducing the rate of drainage. The film and foam are meta-stable because of the high interfacial energy. Also, as for colloidal solutions, van der Waals forces will tend to thin the film.

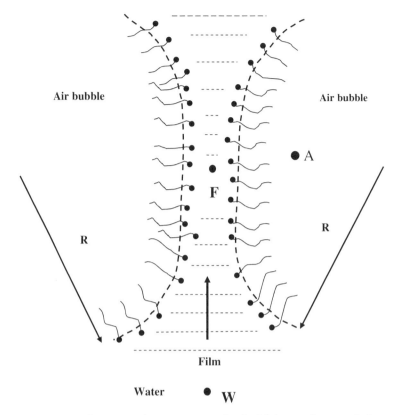

Figure 8.4 Surfactant adsorption at the bubble surface stabilises the films which create a foam.

Thin-film elasticity

In the situation shown in Figure 8.4 we can calculate the pressure difference between points A and W using the Laplace equation: $\nabla P = 2\gamma/R$. Across a flat interface there is no pressure difference, so the pressure in the film (at F) must be greater (by ∇P) than the pressure in the bulk liquid (at W). The pressure increase in the film is due entirely to the repulsive interaction between the two surfaces, which may be of electrostatic origin due to the ionization of the surfactant head groups. In addition to this increased pressure in the film there is also a second factor which helps to give local stability to the foam. This is basically to do with the response of the film to mechanical shocks and thermal fluctuations, which of course must be withstood by a stable foam. Let us see what happens if a soap film is suddenly stretched during such a fluctuation, as illustrated in Figure 8.5. A region must be instanta-

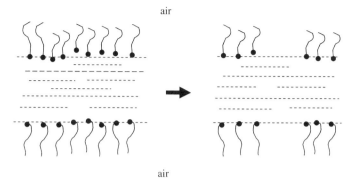

Figure 8.5 The effect of instantaneous stretching of a soap film.

neously produced with exposed pure water interface, before surfactant molecules have time to diffuse from bulk to the surface. Since the surface energy of the water region is much higher than of the surfactant region there will be a strong surface-restoring force to close up the exposed region. This effect gives rise to an apparent elasticity in the film which enhances mechanical stability.

A further elasticity factor arises directly from the properties of the surfactant film itself. Surface diffusion of adsorbed surfactant is much faster than diffusion from bulk (dilute) solution and from the data obtained in Langmuir–Blodgett studies, later in this chapter, increasing the area per head group across the entire film surface, via stretching of the film, will substantially increase the surface tension of the entire film, when the surfactant layer is initially in the fully packed or compressed state. This nonlocal effect generates a substantial elasticity in the film given by $\partial\gamma/\partial A$ and protects the film from mechanical disturbances. An understanding of the mechanisms involved in producing stable foams leads directly to the methods used to destroy them. Surfactant foams can be destroyed simply by spraying with ethanol, which lowers the surface tension and allows the surface film to grow via its rapid diffusion from bulk to surface.

Repulsive forces in thin liquid films

The equilibrium thickness of a (meta-)stable soap film will depend on the strength and range of the repulsive forces in the film. Electrostatic forces are long-range in water and hence give rise to thick (0.2 micron) films, which are highly coloured due to the interference of visible light

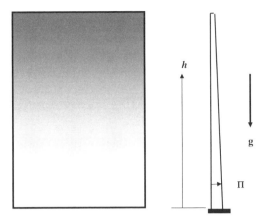

Figure 8.6 Schematic diagram of the effect of drainage under gravity on the thickness of a large soap film.

reflected from the film (it is for the same reason that oil films spilt on the road often show a variety of colours). Films can also be stabilized by very short-range (<5 nm) steric and solvation forces. These are called 'Newton black films' because visible light reflected from the front and back surfaces destructively interferes and so the films appear black. A beautiful range of colours can be observed by carefully allowing a large soap film, held on a rectangular metal frame, to slowly drain under gravity (see Figure 8.6).

In this situation, the equilibrium thickness at any given height h is determined by the balance between the hydrostatic pressure in the liquid ($h\rho g$) and the repulsive pressure in the film, that is: $\pi = h\rho g$. Cyril Isenberg gives many beautiful pictures of soap films of different geometries in his book *The Science of Soap Films and Soap Bubbles* (1992). Sir Isaac *Newton* published his observations of the colours of soap bubbles in *Opticks* (1730). This experimental set-up has been used to measure the interaction force between surfactant surfaces, as a function of separation distance or film thickness. These forces are important in stabilizing surfactant lamellar phases and in cell–cell interactions, as well as in colloidal interactions generally.

Froth flotation

Materials mined from the Earth's crust are usually highly heterogeneous mixtures of amorphous and crystalline solid phases. Crushing and grinding operations are employed to liberate individual pure grains in the 10–50 μm size range. One of the most widely used (10^9 tons per year) processes for separating the required mineral from the 'gangue' (unwanted material, e.g. quartz) exploits the wetting properties of the surface of the grains in a froth flotation process. This cheap and relatively simple process is based on the observation that hydrophobic (or non-wetting) particles dispersed in water easily attach to air bubbles which carry them upwards to the top of the flotation chamber where they can be collected. Hydrophilic particles do not attach to the bubbles and remain dispersed in bulk solution This important industrial process was invented in Australia by Charles Potter, a Melbourne brewer, in 1901. The first commercial process was set up in Broken Hill.

The common gangue material quartz (silica) is naturally hydrophilic and can be easily separated in this way from hydrophobic materials such as talc, molybdenite, metal sulphides and some types of coal. Minerals which are hydrophilic can usually be made hydrophobic by adding surfactant (referred to as an 'activator') to the solution which selectively adsorbs on the required grains. For example, cationic surfactants (e.g. CTAB) will adsorb onto most negatively charged surfaces whereas anionic surfactants (e.g. SDS) will not. Optimum flotation conditions are usually obtained by experiment using a model test cell called a 'Hallimond tube'. In addition to activator compounds, frothers which are also surfactants are added to stabilize the foam produced at the top of the flotation chamber. Mixtures of non-ionic and ionic surfactant molecules make the best frothers. As examples of the remarkable efficiency of the process, only 45 g of collector and 35 g of frother are required to float 1 ton of quartz and only 30 g of collector will separate 3 tons of sulphide ore.

The flotation process is used in the early separation stage for obtaining pure metals from Cu, Pb, Zn, Co, Ni, Mo, Au and Sb ores, whether sulphides, oxides or carbonates. It is also used to concentrate CaF_2, $BaSO_4$, NaCl, KC1, S, alumina, silica and clay. Ground coal is also treated to remove ash-producing shale, rock and metal sulphides which cause air pollution by SO_2 during combustion. In recycling processes, ink can be removed from paper and metallic silver from photographic residues, using flotation. It is even used for removing earth from vegetables during cleaning.

The Langmuir trough

Surfactant molecules will adsorb onto a wide range of solid substrates from aqueous solution. The amount and type of adsorption depends

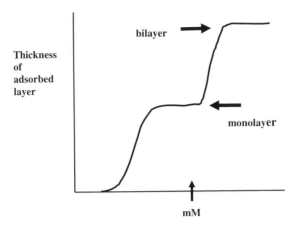

Figure 8.7 Schematic diagram of simple monolayer and bilayer surfactant adsorption from aqueous solution.

on the solution concentration, the nature of the surfactant and the substrate. Generally, ionic surfactants adsorb on oppositely charged surfaces and non-ionic ones adsorb on most surfaces. Often, monolayer and bilayer adsorption is observed, as shown in Figure 8.7. In this example, the cationic surfactant CTAB adsorbs on mica (a negatively charged aluminosilicate) with a well-defined step adsorption isotherm. A monolayer usually adsorbs well below the cmc (in this case around 1 mM), whereas a bilayer forms only at the cmc. Since the mica substrate is hydrophilic we would expect the water contact angle to be a maximum for the hydrophobic monolayer and then fall when the bilayer adsorbs, as has been reported. These observations correlate well with the flotation behaviour of the mineral. Clearly, surfactant adsorption can completely alter the wetting properties of a substrate surface and can also be used to control the coagulation of sols via their effect on the particle's charge.

In the case of adsorption from solution, the surfactant layers are in equilibrium with the solution and will de-sorb on dilution. However, it would be very useful to produce adsorbed layers in both air and water, which will remain adsorbed. This can be achieved using the Langmuir–Blodgett deposition technique. The technique is based on the observation that if a surfactant, which is insoluble in water, is dissolved in a volatile, non-aqueous solvent and then spread on water, an insoluble monolayer of orientated surfactant molecules will remain at the air/solution interface. The effect of the spreading surfactant and its surface film pressure can be dramatically demonstrated by spreading hydrophobic talc powder on a clean water surface and then placing a droplet of surfactant solution gently in the middle. The talc is rapidly displaced in all directions from the centre by the spreading surfactant. This is the basis of the 'camphor duck' toy, which 'swims' around a water bath via dissolution and surface spreading of a small piece of camphor attached to its tail.

The behaviour of spreading films on water and the old observation of 'pouring oil on troubled waters' has a long history. Plinius the Elder, who died in the Mt Vesuvius eruption of AD 79, refers to spreading of oil on water, as does the Venerable Bede (673–735). More recently, Benjamin Franklin did his experiments on Clapham Common near London, on the area of water which could be stilled by a spreading film, and his results were transmitted by his friend William Brownrigg to the Royal Society, which promptly published it (with slight modifications) in *The Philosophical Transactions of the Royal Society* (1774). Later, Lord Rayleigh made the important assumption that the spreading film was monomolecular to make the first estimate of molecular size. His estimate was actually quite close to the thickness of a film of oleic acid, the main component in olive oil. Lord Rayleigh's experiments were reported in *Proceedings of the Royal Society* (1890).

Spreading of an insoluble (or temporarily insoluble) surfactant monolayer effectively produces a two-dimensional surface phase. This thin molecular layer exerts a lateral 'film pressure', which can be simply demonstrated by covering a water surface with a uniform layer of finely divided hydrophobic talc and placing a droplet of surfactant solution (0.003M CTAB solution) in its centre. The effect of the film pressure of the spreading surfactant is dramatic, as seen in Figures 8.8 and 8.9.

Figure 8.9 shows talc symmetrically removed from the centre by the application of one droplet of 0.003 M CTAB solution. The process occurs rapidly (in less than 0.1 s), due to the surfactant's rapid rate of surface diffusion. The hydrophobic talc is compressed against the glass walls of the crystallizing dish. Placing more droplets at the glass edges displaces the talc layers, which move like tectonic plates to produce scaled maps of the world, as in Douglas Adams's *Hitchhiker's Guide to the Galaxy*.

We can study these lateral surface forces by measuring their action on a freely movable beam separating the (insoluble) surfactant coated surface from the pure water surface, as illustrated in Figure 8.10. That the force per unit length (F/l) acting on the barrier is given by ($\gamma_w - \gamma_s$) can be easily shown by consideration of the free energy change on allowing the barrier to move an infinitesimal distance to the right; thus,

Figure 8.8 Photograph of hydrophobic talc powder spread uniformly on the surface of water.

$$dG = +ldx\gamma_s - ldx\gamma_w$$

and the force acting on the barrier must be given by the gradient in free energy of the system:

$$F = -dG/dx \quad \text{and hence} \quad F/l = \gamma_w - \gamma_s$$

The force per unit length (F/l) generated by the surfactant film is called the 'surface film pressure' Π_F. In the Langmuir trough device, illustrated in Figure 8.11, the density and hence the pressure in the film can be varied via a movable barrier.

The first description of a Langmuir trough appears to have been given by Langmuir in *Journal of the American Chemical Society* (1848).

Figure 8.9 Instantaneous removal of the talc in the centre caused by a droplet of soap solution.

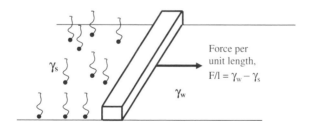

Figure 8.10 Schematic diagram of the forces acting on a beam separating a pure water surface from a surfactant coated surface.

Agnes Pockels carried out many experiments which were reported in *Nature* (1891). This was a remarkable achievement for a woman of the time, working from home. A schematic diagram of a typical experimental set-up is given in Figure 8.12. In the case illustrated, the surface

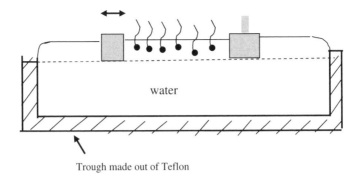

Trough made out of Teflon

Figure 8.11 Schematic sectional diagram of the Langmuir trough showing a trapped film of insoluble surfactant molecules.

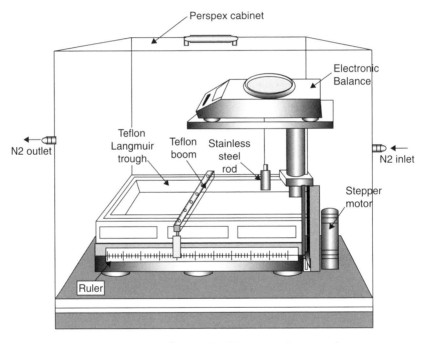

Figure 8.12 Diagram of a typical Langmuir trough apparatus.

tensions are measured using the rod in free surface technique and a motor is used to move the Teflon boom. If a solid substrate is withdrawn vertically through the surfactant side of the Langmuir trough, a uniform monolayer can be transferred to the solid, as illustrated in Figure 8.13.

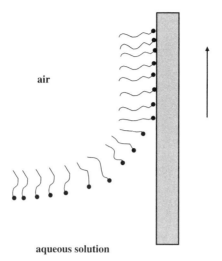

air

aqueous solution

Figure 8.13 Langmuir-Blodgett coating of surfactant monolayers.

Since the density of the monolayer can be varied using the applied surface pressure (via movement of the boom) a wide variety of monolayer conformations can be deposited. Because the surfactant is insoluble, the layer will be stable in both air and aqueous solutions. A second bilayer can be adsorbed on the plate, if it is re-immersed in the trough. The layers adsorbed using this technique are called *Langmuir–Blodgett (LB) layers*. An example of a close-packed surfactant LB coating on a smooth polished substrate, as imaged by an atomic force microscope, is given in Figure 8.14.

Using special surfactants, the coated layers can be polymerized to increase their strength (e.g. C_{11}—C≡C—C≡C—C_7COOH polymerizes on exposure to UV irradiation). These monolayers improve the characteristics of photo-semiconductors. A power increase of 60 per cent can be achieved by the deposition of only two monolayers. LED devices can be improved in efficiency by an order of magnitude by depositing eight surfactant layers, apparently because of a tunnelling mechanism. The alternative method of vacuum deposition of thin layers causes damage to the device because of heating at the surface during evaporation and, in addition, does not produce such uniform layers. LB layers are now also being used for photo-resist masking on silicon. The resolution at present is about a micron but using LB layers it could be reduced to 0.1 micron. Finally, LB films also act as an effective lubrication layer between solids.

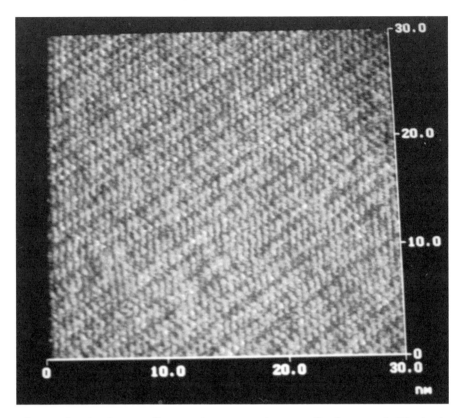

Figure 8.14 Atomic force microscope image of a Langmuir-Blodgelt surfactant monolayer.

Langmuir–Blodgett films

Since the total number of molecules of surfactant added to the Langmuir trough is known (from the molecular weight) the area per molecule is also known and can be varied simply by moving the boom. The relation between the film pressure and the area per molecule can, therefore, be measured. This is in fact a very elegant method for the study of molecular films. The precise isotherm is characteristic of the surfactant but the general features often observed are shown in Figure 8.15.

The various regions on the isotherm are determined by the lateral interaction between the surfactant molecules within the surface phase. In the dilute, 'gaseous' state, the molecules can be considered to be negligible in size and non-interacting. Under these conditions the isotherms obey an ideal, two-dimensional gas equation of the form $\pi A = kT$. As the pressure is increased, a point is reached (at about $8\,\text{nm}^2$ for myris-

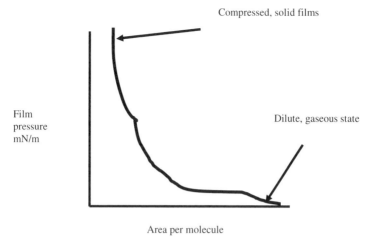

Figure 8.15 Typical film pressure isotherm for a surfactant monolayer.

tic acid) where the attractive forces between the hydrocarbon tails cause a condensation process, analogous to liquid condensation from the vapour phase. However, at the end of this process (at about $0.5\,nm^2$ in the example given) the surface layer is not completely condensed because of the strong, relatively long-range electrostatic repulsion between the ionic head groups. This head-group repulsion keeps the surface layer fluid, whilst the attractive van der Waals forces between the hydrocarbon chains keeps the film coherent. In this state of modified ('real') gas equation can be used to describe the isotherm of the form

$$(\pi - \pi_0)(A - A_0) = kT$$

where π_0 and A_0 are correction terms related to the attractive and repulsive forces between the molecules.

At still higher pressures, the film becomes completely packed (and will eventually buckle) and the limiting area corresponds to the cross-sectional packing area of the surfactant molecule. This region is also interesting because it demonstrates that compressed surface films will respond to even small increases in surface area, such as by stretching a surface through mechanical vibrations, with a large increase in surface energy of the entire film. This behaviour generates a surface elasticity which is important in giving mechanical stability to soap films (discussed earlier) and is also the cause of the 'oil on troubled waters' phenomenon, observed for more than two millennia.

For more information, see the standard text *Insoluble Monolayers at Liquid–Gas Interfaces* by G.L. Gaines (1966).

Experiment 8.1   Flotation of powdered silica

Introduction

Froth flotation is one of the simplest and most widely used separation techniques for mineral ores. Crushing and grinding of highly heterogeneous rocks liberates individual pure grains in the 10–50 μm size range. These particles can often be readily separated using differences in their surface properties. When one of the components is hydrophobic and the other hydrophilic, gas bubbles passed through a stirred aqueous suspension carry the hydrophobic particles to the top of the vessel, where they can be collected. The wetting properties of the surface of the pure mineral determine the flotation efficiency. Thus, if water has a high contact angle on the mineral its flotation efficiency will also be high.

Often the minerals we want to float are hydrophilic, and surfactants (called 'collectors') are added, which, at a specific concentration, adsorb onto the particle surface, making it hydrophobic and hence floatable. In a mixture of hydrophilic minerals, optimum flotation will occur where one of the minerals adsorbs collector but the others do not.

In industrial processes the flotation cells have, of course, a very large capacity. However, the efficiency of the process can be determined on a much smaller scale using a Hallimond apparatus (see Figure 8.16). In this experiment we will use this apparatus to measure the flotation efficiency of hydrophilic Ballotini beads (~50 μm diameter) over a range of CTAB (cetyltrimethylammonium bromide) concentrations. The glass beads (when clean) are naturally hydrophilic because of a high density of surface silanol groups which hydrogen-bond strongly with water. Adsorption of a monolayer of CTAB makes the surface hydrophobic but at higher CTAB concentrations a bilayer adsorbs, which makes the surface again hydrophilic. It is the solution concentration that controls this adsorption and hence the flotation efficiency of glass.

Experimental details

The Ballotini glass beads (~50 μm diameter) used in these experiments should be cleaned in dilute NaOH solution, washed thoroughly with

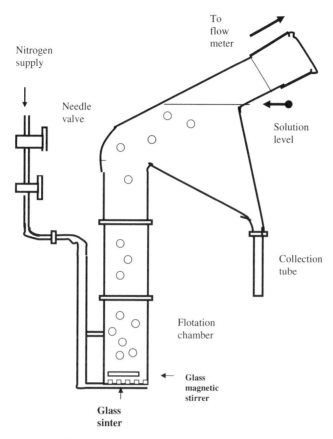

Figure 8.16 Hallimond tube used to measure flotation efficiency.

distilled water and then completely dried. The surface of the powder should then be hydrophilic, with a high density of silanol groups (about 1 per 25Å^2) which makes the particles hydrophilic.

The glass Hallimond tube apparatus typically used to study flotation in the laboratory is illustrated in Figure 8.16.

Instructions

1. Dismantle and clean the flotation apparatus by scrubbing with ethanol followed by rinsing with distilled water.

2. Put the apparatus together as shown in the diagram but leave off the top. Close the tap connecting the gas line to the flotation chamber.

3. Fill with a known volume of distilled water to just below the solution level indicated in the figure. The top of the apparatus must be connected to a suitable low-rate flowmeter. (A soap bubble column is adequate.)

4. Adjust the needle valve until the N_2 gas flow rate is about 50 mL/min. Note this rate.

5. Add distilled water (whilst the gas is still flowing) to bring to the solution level indicated by an arrow in the figure, and note the total solution volume. (At this level particles carried up to the surface will be collected and will not return to the flotation chamber once the bubbles rupture at the solution surface.)

6. Switch off (or bypass) the gas flow such that the flow-rate remains the same on re-opening the flow. Consistency of the flow-rate is necessary to obtain accurate comparisons of flotation efficiency. Check this by re-measuring the flow-rate.

7. Disconnect the N_2 line (at the needle valve on the gas cylinder) and pour out the distilled water.

8. Accurately weigh about 1 g of clean, dry Ballotini powder and place it into the flotation chamber, half-fill with a known volume of distilled water and drop in the glass magnetic stirrer supplied (be careful as the latter is easily broken). Accurately weigh the dry collection tube and then fit onto the apparatus.

9. Stir the powder and solution for 5 min to 'condition' the surface; allow to settle.

10. Reconnect the rest of the apparatus to the flotation chamber and add more distilled water to bring to the already determined total solution volume. Connect to the flow-meter and begin stirring the flotation chamber.

11. Vent the N_2 to atmosphere for a few seconds, if using a bypass system, to remove pressure build-up in the line prior to each run.

12. Allow the N_2 gas to flow for precisely 30 s, making sure that the flow-rate remains the same as before. Switch off the magnetic stirrer and allow 5 min to settle.

13. Pour off excess solution from the apparatus and remove the collection tube. Allow the powder to settle in the tube and then pour off as much excess solution as possible without losing any of the powder and dry at about 120°C for at least 20 min in an oven (loosely cover to prevent dust settling on the powder during drying) or leave overnight. Reweigh the collection tube.

14. Calculate the percentage flotation efficiency, that is: % collected/original wt.

15. Repeat and take the average value.

16. Follow the procedure described above using 1 g portions in 10^{-7}, 10^{-6}, 10^{-5}, 10^{-4} and 10^{-3} M CTAB solutions. Addition of CTAB may cause extensive frothing. Since the powder will be retained in this foam it must be broken up either mechanically using a Pasteur pipette or by the addition of a small amount of ethanol. The powder may also stick to the walls of the apparatus above the collection tube. This powder should be dislodged by gentle tapping on the walls of the apparatus. Calculate the mean flotation efficiency at each concentration and graph the results.

Contact angles

Clean a soda glass plate by washing in 10% NaOH (care) followed by rinsing in double-distilled water. Blow dry with N_2 and observe the behaviour of a droplet of clean water on the plate. Blow dry again, place droplets of the various CTAB solutions used on the clean plate and observe the wetting behaviour with CTAB concentration.

Questions

1. Explain the flotation recovery results obtained with CTAB.

2. What are the major differences between this experiment and a typical industrial flotation process (other than size)?

3. Why did ethanol break up the surfactant foam?

Appendix 1
Useful Information

Fundamental constants

Boltzmann constant, $k = 1.381 \times 10^{-23}\,\mathrm{J\,K^{-1}}$

Electronic charge, $-e = 1.602 \times 10^{-19}\,\mathrm{C}$

Permittivity of free space, $\varepsilon_0 = 8.854 \times 10^{-12}\,\mathrm{C^2\,J^{-1}\,m^{-1}}$

$kT = 4.12 \times 10^{-21}\,\mathrm{J}$ at $298\,\mathrm{K}$

$1\ \mathrm{atm} = 1.013 \times 10^{5}\,\mathrm{N\,m^{-2}}$ (Pa)

$kT/e = 25.7\,\mathrm{mV}$ at $298\,\mathrm{K}$

$1\,\mathrm{C\,m^{-2}} = 1$ unit charge per $0.16\,\mathrm{nm^2}$

Viscosity of water $= 0.001\,\mathrm{N\,s\,m^{-2}}$ at $20°\mathrm{C}$

Viscosity of water $= 0.00089\,\mathrm{N\,s\,m^{-2}}$ at $25°\mathrm{C}$

Dielectric constant of water $= 80.2$ at $20°\mathrm{C}$

Dielectric constant of water $= 78.5$ at $25°\mathrm{C}$

Zeta potentials

From microelectrophoresis measurements on a spherical colloid particle, the observed elctromobility U_E is directly related to the zeta potential by the equation:

Applied Colloid and Surface Chemistry Richard M. Pashley and Marilyn E. Karaman
© 2004 John Wiley & Sons, Ltd.   ISBN 0 470 86882 1 (HB)   0 470 86883 X (PB)

$$\zeta(mV) = 12.8 \times U_E\left(\mu m\,s^{-1}/V\,cm^{-1}\right)$$

$$\zeta(mV) = 1.28 \times 10^9 \times U_E\left(m\,s^{-1}/V\,m^{-1}\right) \text{ at } 25°C$$

Debye lengths

$$\kappa^{-1}(nm) = \frac{1.0586 \times 10^{13}}{\left[\Sigma_i C_i(B)Z_i^2\right]^{1/2}} \quad \text{at } 21°C,\ C_i(B) \text{ in nos/m}^3$$

$$\kappa^{-1}(nm) = \frac{0.305}{\sqrt{M}} \qquad \text{for 1:1 electrolytes, where } M \text{ is mol per L at } 21°C$$

Surface charge density

The surface charge density on a flat, isolated surface immersed in 1:1 aqueous electrolyte solution, at 25°C, is given by the relation:

$$\sigma_0(Cm^{-2}) = \frac{3.571 \times 10^{-11}}{\left(\kappa^{-1}/m\right)} \times \sinh(\psi_0/mV \times 0.01946)$$

Appendix 2
Mathematical notes on the Poisson–Boltzmann equation

The solution to (6.12) in Chapter 6: $d^2Y/dX^2 = \sinh Y$ depends upon the boundary conditions of the system under consideration. The two main cases of interest are: (a) an isolated surface and (b) interacting surfaces. In both cases we can carry out the first integral of (6.12) using the identity

$$\int \frac{d^2y}{dx^2} dy = 1/2 \left(\frac{dy}{dx}\right)^2 + C$$

and hence

$$\left(\frac{dY}{dX}\right)^2 = 2\cosh Y + C \qquad\qquad (A.1)$$

Determination of the integration constant C depends on the boundary conditions. For case (a) an isolated surface (Figure A.1), $\left(\dfrac{dY}{dX}\right) = 0$ when $Y = 0$ and hence $C = -2$ and (A.1) becomes

---

Applied Colloid and Surface Chemistry Richard M. Pashley and Marilyn E. Karaman
© 2004 John Wiley & Sons, Ltd.   ISBN 0 470 86882 1 (HB)   0 470 86883 X (PB)

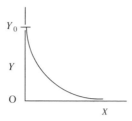

Figure A.1 Potential decay away from a flat surface.

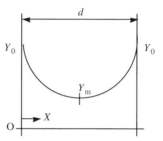

Figure A.2 Interaction between two (flat) electrical double-layers.

$$\left(\frac{dY}{dX}\right)^2 = 2(\cosh Y - 1) \tag{A.2}$$

This can be integrated to obtain an exact analytical equation for the potential decay away from a flat surface (see Chapter 6).

The simplest example of case (b) is for the interaction of identical surfaces, i.e. the symmetrical case shown in Figure A.2. In this case we can use the boundary condition $(dY/dX) = 0$ when $Y = Y_m$ (the mid-plane potential) and hence $C = -2 \cosh Y_m$. Equation (A.1) then becomes

$$\left(\frac{dY}{dX}\right)^2 = 2(\cosh Y - \cosh Y_m) \tag{A.3}$$

Unfortunately, integration of this equation is rather difficult and leads to elliptic integrals which only have numerical solutions. (A relatively simple numerical solution to (A.3), without the use of elliptic integrals, was developed by Chan et al., 1980.)

Even though this equation is difficult to solve, many approximate methods have been used. Equation (A.3) is, however, interesting for what it tells us about the double-layer interaction. It can be rearranged in the form

$$\left(\frac{dY}{dX}\right)^2 - 2\cosh Y = 2\cosh Y_m \neq f(Y \text{ or } X) \tag{A.4}$$

Clearly, the sum of the two terms on the left-hand side is constant at any point between the two interacting surfaces.

We can now identify the first term in (A.4) with Maxwell's stress tensor, which acts on any dielectric in an electric field. The magnitude of this force $|F_E|$ is given by

$$|F_E| = \frac{\varepsilon_o D|E|^2}{2} \tag{A.5}$$

where the electric field strength $|E|$ is proportional to dY/dX. This electrostatic force acts on each surface to pull it towards its oppositely charged diffuse layer. The second force is due to the osmotic pressure generated by the excess of (charged) solute counter-ions in the interlayer between the interacting surfaces, compared with the bulk solution. This pressure acts to push the surfaces apart and is proportional to the (total) local electrolyte concentration. It can be easily demonstrated that the local electrolyte concentration is in turn proportional to $\cosh Y$, which is actually the sum of the Boltzmann factors for each ion.

It follows that for interacting flat surfaces at any given separation distance, two opposing forces – electrostatic and osmotic – vary in magnitude across the liquid film but compensate each other to give the overall repulsive pressure.

For the symmetrical case it is easier to obtain the value of this pressure using the fact that at the mid-plane, where $Y = Y_m$, the electrostatic term disappears (i.e. $dY/dX = 0$, $|E| = 0$) and the total pressure is equal to the osmotic pressure at this plane, which can then be easily calculated once the value of Y_m is known.

Appendix 3
Notes on Three-dimensional Differential Calculus and the Fundamental Equations of Electrostatics

Div: $\quad \vec{\nabla} \bullet \vec{E}(\vec{r}) = \dfrac{\rho(\vec{r})}{\varepsilon_0 D} = \dfrac{\partial E_x}{\partial x} + \dfrac{\partial E_y}{\partial y} + \dfrac{\partial E_z}{\partial z}$ (a scalar)

Curl: $\quad \vec{\nabla} \times \vec{E}(\vec{r}) = 0$

Grad: $\quad \vec{E}(\vec{r}) = -\vec{\nabla}\psi(\vec{r}) \equiv -\left(\dfrac{\partial \psi}{\partial x}, \dfrac{\partial \psi}{\partial y}, \dfrac{\partial \psi}{\partial z} \right),$ i.e. the vector i_x, i_y, i_z

$$\vec{E}(\vec{r}) = \lim_{q=0} \dfrac{\vec{F}(\vec{r})}{q}$$

Gauss's law: $\quad \oint\limits_{surface} \vec{E} \bullet \hat{n}\mathrm{d}a = \dfrac{\sum\limits_{i}^{\text{insurface}} q_i}{\varepsilon_0}$

Applied Colloid and Surface Chemistry Richard M. Pashley and Marilyn E. Karaman
© 2004 John Wiley & Sons, Ltd. ISBN 0 470 86882 1 (HB) 0 470 86883 X (PB)

Index

Note: Figures and Tables are indicated by *italic page numbers*

Applied Colloid and Surface Chemistry Richard M. Pashley and Marilyn E. Karaman
© 2004 John Wiley & Sons, Ltd.   ISBN 0 470 86882 1 (HB)   0 470 86883 X (PB)

Index compiled by Paul Nash